# Imperial China
## A Beginner's Guide

T0072122

**ONEWORLD BEGINNER'S GUIDES** combine an original, inventive, and [...] approach with expert analysis on subjects ranging from art and history to religion [...] politics, and everything in-between. Innovative and affordable, books in the series are perfect for anyone curious about the way the world works and the big ideas of our time.

# Imperial China
## A Beginner's Guide

Peter Lorge

**ONEWORLD**

A Oneworld Book

First published by Oneworld Publications in 2021

ISBN 978-1-78607-578-9
eISBN 978-1-78607-579-6

Typeset by Geethik Technologies
Printed and bound in Great Britain by Clays Ltd, Elcograf S.p.A.

Oneworld Publications
10 Bloomsbury Street
London WC1B 3SR
England

Stay up to date with the latest books,
special offers, and exclusive content from
Oneworld with our newsletter

Sign up on our website
**oneworld-publications.com**

MIX
Paper from
responsible sources
FSC® C018072

*This book is dedicated to the memory of my father,
Bernd Lorge (1933–2016), who I hope would have appreciated
my efforts to speak to a larger audience*

# Contents

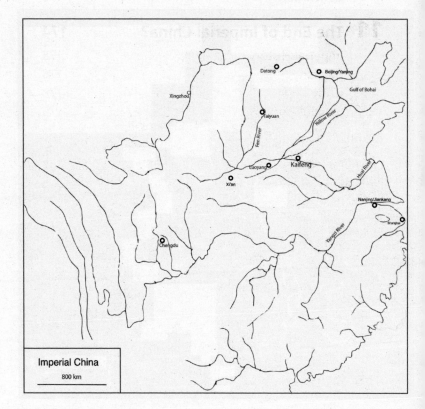

Imperial China

800 km

# A Timeline of the Dynasties of Imperial China

| | |
|---|---|
| Qin | 221–206 BCE |
| Han | 202 BCE–220 CE |
| Former/Western Han | 202 BCE–23 CE |
| Xin (interregnum) | 9–23 CE |
| Later/Eastern Han | 25–220 CE |
| Wei, Jin, Nan-Bei Chao | 220–589 |
| Three Kingdoms | 220–280 |
| Wei | 220–265 |
| Han (Shu Han) | 221–263 |
| Wu | 222–280 |
| Jin | 265–420 |
| Western Jin | 265–316 |
| Eastern Jin | 317–420 |
| Six Dynasties | 222–589 |
| Sixteen Kingdoms | 304–439 |
| Northern and Southern Dynasties (Nan-Bei Chao) | 420–589 |
| Southern Dynasties | 420–579 |
| Liu Song | 420–479 |
| Qi | 479–502 |
| Liang | 502–557 |
| Chen | 557–589 |
| Northern Dynasties | 386–581 |
| Northern Wei | 386–534 |
| Eastern Wei | 534–550 |

| | |
|---|---|
| Western Wei | 535–556 |
| Northern Qi | 550–577 |
| Northern Zhou | 557–581 |
| Sui | 581–618 |
| Tang | 618–907 |
| Five Dynasties and Ten Kingdoms period | 907–960 |
| Northern Dynasties | 907–960 |
| Ten Kingdoms | 907–979 |
| Song | 960–1279 |
| Liao (Kitan) | 916–1125 |
| Jin (Jurchen) | 1115–1234 |
| Yuan (Mongol) | 1279–1368 |
| Ming | 1368–1644 |
| Qing (Manchu) | 1644–1912 |

was clear to many of those early missionaries that it would be very difficult to convert the educated elite. Chinese education was closely tied to its own classics and its own ideology. Until well into the nineteenth century Chinese elites were confident of the superiority of their own culture. For most of imperial history, China had been open to outside influences without feeling compromised by foreign cultures. It was only in the middle of the nineteenth century that Chinese attitudes toward Chinese and foreign, really European, culture began to change. That the imperial house of the last dynasty, the Qing (1644–1912), was Manchu, not Chinese, only served to complicate things further.

Imperial China is therefore very hard to understand when seen only from the perspective of the late Qing dynasty, after Europe had finally surged ahead of the rest of the world technologically. Qing elites, whether Chinese or Manchu, saw themselves as the inheritors of a great civilization. Western technology was interesting and even useful, but Western culture and religion were far less attractive. For their part, Westerners were baffled by the resistance in the nineteenth century to what they believed was an obviously superior culture. Modern technology proved the value of Western culture more generally and, retrospectively, the "correctness" of its path of historical development. In Western eyes, China was backward in the nineteenth century because it failed to follow the Western course of development. Indeed, China was imagined to have been static over the millennia, fixed in its conservativism. There is a significant danger in writing a history of imperial China that generalizations over two millennia emphasize its culture's ideal or normative image of itself, or at least an outsider's view. In the case of China, this is particularly ironic given that there is more history here than anywhere else.

Any introduction to Chinese history in a Western language, therefore, immediately confronts the place of Chinese history in the Western historical imagination. A modern reader inherits a broad range of characterizations of China that are so profound that it seems as if Chinese history cannot be discussed without some reference to those biases and their origins. Nevertheless, Chinese history existed outside the West and, even though many modern Chinese thinkers were themselves obsessed with comparison to the West, it can be discussed without it.

That said, imperial China's encounter, even collision, with the West in the nineteenth and twentieth centuries is a formidable obstacle to understanding Chinese history on its own terms. Qing historian Liang Qichao (1873–1929) actually lamented the fact that "our country has no name," in contrast to modern European nation states. It was only when Qing dynasty officials went abroad that they were informed that their state was called "China." "China" was not, at that time, a translation of the Chinese term "*Zhongguo*" as it is in modern Chinese, or of the Manchu term "*Dulimbai Gurun*." The China the West spoke of was a poorly delineated territory, people, and culture which, at a great distance, appeared unitary, coherent, and ancient. Yet for a brief introductory work covering such a vast period and encompassing many different peoples, territories, and cultures it seems the simplest course of action is to accept the broad terms "China" and "Chinese."

There were two pre-modern time periods in that simplification, the ancient, or pre-imperial period, and the imperial period. Western constructions of China also imposed two misleading concepts on Chinese culture, the creation of Confucianism, and the myth that Chinese characters represented ideas. All of these considerations require some discussion before engaging in the topical issues of imperial Chinese history.

## Before the Imperial Age

Pre-imperial China is often called China's "Classical Age" or its "Axial Age" in order to allow a period equivalent to the Western Classical Age, as well as other ancient cultures. The great foundational thinkers and culture heroes, real or mythical, were purported to have lived in this period, stretching anywhere from the legendary Yellow Emperor (third millennium BCE) or more historical periods like the Spring and Autumn Period (771–476 BCE) and the Warring States period (475–221 BCE), to the founding of the Qin dynasty in 221 BCE. More narrowly, the Classical Age was the time that included the thinkers who came to be seen as foundational to educated culture. For example, men like Master Kong (551–479 BCE)—known as Confucius, founder of Confucianism, in the West—who argued that morality was more important than the law; the legendary Master Lao (Lao Tzu, Laozi), founder of Daoism/Taoism, who was purported to have lived in the sixth century; or Master Mo (c.470–c.391 BCE). Regardless of whether they were merely legendary, or of the reliability of the texts connected to these men, their place in Chinese history became fixed in imperial times and their importance has carried through to the present. Indeed, some people have focused on the Classical Age in China as a way to simplify discussions about Chinese culture in much the same way that Western civilization has sometimes been simplified down to the Ancient Greek thinkers, with some reference to the Romans.

Much of what came to be known about the pre-imperial age was formed during the early Han dynasty (202 BCE–220 CE). Until modern archaeology, with a few notable exceptions, almost every extant text from pre-imperial times was actually formulated, compiled, commented upon, and revised during the Han dynasty. This was not simply an outgrowth of the efforts to recover works destroyed during the Qin dynasty, it was also part of the larger process of formalizing and fixing previously fluid texts, many of which had circulated in fragmentary form. In this sense, Han dynasty scholars constructed a pre-imperial past that justified their imperial present. While there was interest in finding authentic early works, there was, in practical terms, no real way to determine what was or was not genuine. The great debates over which versions of earlier texts were authentic were never really resolved.

The focus on the Classical Age as the true font of Chinese culture served also to suggest that there was an original, pure culture that developed in a place and time which could be returned to, at least textually, and that pre-imperial China was a hermetically sealed culture untouched by outside influences. Of course, this was never true. Sitting on the eastern edge of Eurasia, China has always been in contact with the rest of the continent, as well as the maritime territories of east and southeast Asia. Ideas and goods flowed in all directions, circulating and returning in new forms, giving rise to new discoveries, and enriching all.

For later Chinese, imperial and modern, the only disconcerting thing about the Classical Age was its disunity. The foundations of Chinese culture were laid in a multi-state environment, ruled over by the Zhou king who had spiritual authority over those states. If it produced great thinkers, it also produced great political and social turmoil, leading many to conclude that less diverse thinking might yield a more stable environment. The conflict between diversity and conformity created real tension in imperial China. The "Hundred Schools of Thought" of the

## THE WARRING STATES PERIOD

The Warring States period immediately followed the Spring and Autumn Period, and came to an end with the Qin conquest of all China in 221 BCE. It has been variously dated as beginning in 481, 475, or 403 BCE. During this time the Zhou king was still nominally ruler over the various feudal states of China, but he had no real authority. Instead, there were seven large states involved in the struggle for power. The period takes its name from the *Strategies of the Warring States* (*Zhanguo Ce*), an early text by several hands that records anecdotes and military events from the fifth to the third centuries, just as the Spring and Autumn Period takes its name from the *Spring and Autumn Annals* (*Chunqiu*), a chronicle of the State of Lu (in modern Shandong), and recounts in annalistic form the events from 722 to 481 BCE.

Warring States period stimulated the search for better means of governance and self-cultivation without which the great wisdom of the sages would not have been known. But if some of those thinkers were indeed sages, they needed to be identified and their wisdom followed by everyone. Unity and peace would be created by a shared, common intellectual framework.

The fragility and limited distribution of texts eliminated the work of many thinkers without regard to the qualities of their writing. Oral transmission of teachings was similarly uncertain, and it may well have been that few teachers had any expectation of transmitting their lessons beyond their immediate students. On the other hand, the audience for philosophical thought and written texts was fairly small, confined to the educated upper classes. Some of these men, particularly the knightly class (*shi*) below the actual ruling feudal lords, circulated among the various feudal domains seeking employment and influence. The more compelling teachers were probably more popular, leading to their work spreading more widely and having a higher likelihood of being preserved. It was also true that various stories and teachings were attached to famous teachers in order to give them legitimacy. Consequently, it was often impossible to separate what someone actually said or wrote from what was attributed to them.

In addition to the writings of these teachers, or roving persuaders, the pre-imperial period also bequeathed an extensive set of heroes and myths to Chinese culture. Many of these heroes and myths were promoted by the wandering teachers, either because they believed them to be true and real, or because they served their rhetorical purposes. Mythical rulers like the Yellow Emperor blended into reality with idealized actual rulers, like the founders of the Zhou dynasty, in narratives of correct kingly behavior. In retrospect, it is hard to see much difference between purely mythical figures who exemplified particular behaviors or accomplishments, and idealized historical figures who served the same purpose.

The culture of imperial China rested on a classical tradition of texts and legends that were transformed through evolution, interaction with local, regional, and foreign cultures, and historical happenstance. Governments cared deeply about politics and political ideology, and this led to a general emphasis on the teachings of Master Kong and officials identifying themselves with the Ru (sometimes translated as "classicist"). The importance of Master Kong and other classical texts for government service meant that even as the spoken and written languages changed across China, each dynasty produced and sought to control a central, written culture. That is also why late-nineteenth-century statesmen and foreigners took such a strong interest in trying to establish their view of imperial China, Master Kong, and the Chinese language.

## Imperial China

"Imperial China" is an inherently political term that assumes a great many things about war, ideology, and the construction of the notion of China itself. Most of the documentary record of Chinese history was written by educated men who either worked for the government or aspired to do so. Certainly, very few actively opposed the concept of imperial governance, or the notion that All-Under-Heaven (*tianxia*) should be ruled by a single emperor. Yet, even as these men wrote with this ideology in mind, they also understood that reality did not necessarily coincide with that ideal. There were inherent tensions between morality and power, despite what many thinkers asserted. Morality did not, unfortunately, always or even often compel obedience, and imperial power required the direct application of force. For an emperor to rule a dynasty that claimed authority over All-Under-Heaven was to speak of moral legitimacy while applying force, in the forms of armies, the judiciary, taxes, and required labor service.

It is impossible to define clearly what "China" was in terms of a static territory with a fixed culture and population. Two millennia of historians represented that constantly changing cluster of land, people, and culture as a coherent unit. However, most Chinese scholars knew that there was a deep tension between what remained the same and what changed. The deep commitment to history that has been so often noted in Chinese culture is partly the product of the need to reconcile the abstract constants of cultural ideals with the obvious changes of daily life.

A compromise term for the imprecisely defined territory in which the majority of people would have identified themselves as Chinese is the "Chinese ecumene." This is, admittedly, only a slight improvement over the slightly more old-fashioned term "China proper." The alternative to any such compromise term in discussing this unclear territory over more than two millennia of history is to bog down in endless hedging and diversion. The boundaries of that land, and the boundaries of what constituted Chinese culture, are much clearer in the textual record than they were on the ground in daily life. It is similarly unclear how most of the population in the Chinese ecumene would have defined themselves. Cultural differences for most of imperial history were more obvious than fundamental claims to a shared Chinese identity. Nationalism would not appear in its modern form until the nineteenth century, and then only among some educated men familiar with Western culture. Fundamentally, it is difficult to use a vaguely defined population to lay claim to a vaguely defined territory.

Modern China's territorial claims closely align with the boundaries of the Qing dynasty rather than some clearly established and well-known place called China. This highlights the significant political problem of defining imperial China's geography. All Chinese governments made territorial claims concerning the physical extent of their authority. Those claims were based partly on history and partly on the operational reach of a government's armed forces. This would be a difficult historical question by itself, but the implications of imperial Chinese claims to territory have very real twenty-first century ramifications.

Territory and ethnicity were linked in political and cultural claims with respect to the Chinese ecumene. The fluctuating boundaries and authority of imperial governments, however, were never confined to only Chinese people. Steppe groups were brought into the territories of many dynasties, and other non-Chinese people, however they were understood to stand apart, were always claimed (along with the territory in which they lived) as subjects by central imperial governments. The steppe, a vast grassland that stretched from northeast Eurasia, north of China, to the edge of Europe, was home to a complex and changing group of peoples who moved across Central Eurasia, usually on horseback, butting up against sedentary empires to their south. Imperial Chinese history encompasses a series of empires that incorporated territory with widely disparate populations and cultures. Imperial dynasties were not confined to a single people, even if many of the identifiable minority groups were routinely oppressed by the central government. Like any empire, imperial Chinese governments claimed all the lands and peoples living on those lands as their natural territory and subjects. Indeed, they all claimed natural authority over more than they ever actually controlled.

## Master Kong, the Ru, and Confucius

The translation of Chinese culture for the Western audience has always been subject to more significant biases than the translation of Chinese language. One of the most significant and persistent biases was the construction of the teachings of the Ru as "Confucianism." Confucianism was not only a Western construction; it was a Chinese construction as well. In the nineteenth century, Western missionaries wanted to find a way to convert educated Chinese people by convincing them that Christianity and Ruist teachings (Classical

Chinese education and morals exemplified by Master Kong) were essentially the same, excepting the absence of Jesus Christ from the latter. Ruist teaching was so similar, in fact, that it was almost as if the Chinese had simply lost its religious aspects at some time in the Classical past. The great Ruist sage Master Kong, or Confucius, filled the role that Jesus played in Christianity. Confucianism became a secular philosophy that only needed the readmission of religion in the form of Jesus to return to its true roots.

In China, a different motivation for promoting Confucianism animated men like Kang Youwei (1858–1927). Although he was well aware that Ruism was not Confucianism in the past, Kang believed that a new religion like Christianity was necessary to bind together Chinese people as the Qing dynasty collapsed. For Kang, Confucianism would aid in the creation of a nation distinct from the imperial government. Just as Christianity held together the West and made it powerful, so too would Confucianism hold together China and allow it to modernize.

The Confucianism of the nineteenth century was not just a simplified cultural translation of past practice, it was a purposeful re-conception of Ruist thought, Master Kong, and the place of religion in Chinese society. Most obviously, religion was much more important and much more embedded in Chinese culture than the formulation of a secular Confucianism suggested. China has often been portrayed as a secular state in contrast to the West's religiosity. Some of this secularization predated the invention of Confucianism, with many Europeans intrigued by the idea of civil service exams testing education received outside the Church. Of course, that education was only secular if the Ruist texts were secular, and so the aspects of those works that seemed to invoke some sort of religious belief were downplayed. Buddhism and Daoism were the only religions in China. Confucianism was the rational aspect of an educated man's life (and, in keeping with almost everywhere else at the time, it was only men), which he separated from any religious inclinations.

Confucianism also allowed Westerners to conflate Master Kong's emphasis on morality over law with a general lack of law in China. This dovetailed with Westerners' ignorance of Chinese legal practice to become another point of difference between the two worlds. These comparative ideas of the origins and roles of law framed China as a deficient version of the West and sought fundamental cultural reasons for this. Law, in this telling, was simply an instrument of the state, and had no spiritual or religious origin. Thus, Confucian ideas of the state were substituted for actual knowledge of law in Chinese government.

For most of imperial history, the learning of the Ru was an intellectual tradition, and Master Kong was its greatest thinker. It had aspects that could be regarded as religious, and it played varying roles for educated men. Like any intellectual tradition in competition with others, such as Daoism and Buddhism, those supporting it used every possible argument against its competitors. The late-imperial Chinese and Western reading of Ruism as Confucianism was just that, a late-imperial view of the broad tradition of Ru learning. Just as Ruist thought underwent major shifts in emphasis and interpretation over two millennia, like those of the great thinkers Zhu Xi (1130–1200) in the Song, and Wang Yangming (1472–1529) in the Ming, so too did the Chinese encounter with the West add a new interpretive spin in the nineteenth century. The very fact that Ruist thought could accommodate such major shifts without collapsing shows its depth, breadth, and flexibility.

## Language

If the biases of cultural translation were more significant than those of language translation, it is not to say that the difficulties of translating Chinese were and are insignificant. The written language of pre-imperial China is usually called Classical or Ancient Chinese (*guwen*), and the written language of imperial China,

Literary Chinese (*wenyanwen*). This simple distinction obfuscates many nuances of language and suggests, incorrectly, that Literary Chinese remained the same for two millennia. Spoken languages were even more diverse. Many varieties of Sinitic or Chinese were and are unintelligible to speakers of other varieties of Chinese, making them distinct languages rather than dialects. The written language in imperial times, by contrast, has always been intelligible among the educated, no matter where they grew up or lived in China—or outside it.

Classical Chinese reflected both the spoken and written conventions of a more monosyllabic language. Because everyone learning to read and write in imperial China started with the "Classic" texts written in Classical Chinese, every educated person was well acquainted with what would have been an increasingly archaic language. As spoken Chinese and written Chinese changed and became more polysyllabic, both deviated more and more from the language of the classical texts. Literary Chinese also changed significantly over time, and was not entirely in sync with changes in how Chinese was spoken. While spoken language varied across the Chinese ecumene, the written language maintained more consistency, partly because of education in the classics and partly because of imperial governments.

Literacy was a skill for government and a marker of status. Keeping in mind that books were rare and expensive until at least the eleventh century, both literacy and textual knowledge would have been difficult to attain. Bureaucracy had developed well before the imperial age, and literate officials were necessary to all functions of government. Of course, literacy had many gradations, and the knowledge of classical texts possessed by government officials was not required by the larger number of copyists and recorders of basic information. The well-educated were also expected to be able to write poetry and prose in specific styles (see Chapter 8). Gentlemen were taught that the correct and highest realization of their learning was government service.

However, as not only literacy but education became more wide-spread after the eleventh century, without a concomitant increase in government jobs, the connection loosened. Classical and literary learning became a marker of membership in the literati, or gentry class, with only a few of the luckiest men passing the civil service exams and entering government service.

Education was ideologically oriented toward a unified, empire-wide standard established by the central government. Regionalism or localism were therefore always contrary to imperial pretensions, in the written language if not in the spoken. Historians in the Han dynasty constructed a break between themselves and the pre-imperial past by introducing a new script with a unified set of characters. This unification of characters supposedly accompanied the raft of other unifications of the preceding Qin dynasty, including axle width, weights and measures, and thought. As part of creating the first unified empire, the Qin had necessarily eliminated the cultural variations, in every aspect of life, that existed during the Warring States period. Each state's script variations, histories, and material cultures had to be destroyed. Yet this sudden imperial shift was an artifact of the Han historians. The various regional character variations had begun to converge before the Qin conquest, and it would take a century or more for all of the variations to finally disappear.

## Conclusion

China changed dramatically over the two millennia of imperial Chinese history. That change was so great that any generalization is certain to be wrong at many times and in many places. At the same time, there was a constant political and intellectual argument for the coherence and consistency of something that could be called "China." Even as the boundaries of different dynasties shifted, the literati were taught and continued to believe that they were the inheritors of a powerful, rich culture. Imperial governments and rulers sought to shape that culture for their own ends, and to claim authority over it. The continuous struggle for control of Chinese culture, and temporal authority over territory make simple definitions impossible.

There were, nonetheless, some consistent cultural forms present in all imperial dynasties. The most basic was that there should be a unified, central government ruled by an emperor who possessed Heaven's Mandate. That emperor was Heaven's Son, making him both a temporal ruler over All-Under-Heaven, and a divinely approved ruler. Imperial ritual had real and symbolic power, with a good ruler obtaining divine support for his state, and a bad one courting misfortune. Good fortune, whether political or military, was a sign of divine support. Success was proof of the right to rule.

Imperial China existed mostly in the minds of the elites. The average farmer knew very little about the changes in government or high culture, and had nothing at all to do with "national" issues. Imperial China did not produce a modern nation-state, but reformers in the nineteenth century worked very hard to transform the Chinese people into citizens, and the Qing empire into a country. However, the Qing empire collapsed before those efforts bore fruit. Imperial China as it had existed for two millennia did not or could not make the transition into the modern world.

Retrospectively, the imperial model was a dead end. Yet very few educated people before the nineteenth century would have questioned it. They might have lamented the people running it, or their own lack of success within it, but the ideal was still an emperor ruling a unified empire. Imperial China was never static, though as a highly developed and literate culture it tended self-consciously to look backward. History always mattered, and ignorance of that past was never admirable. Yet until the fall of the Qing dynasty, there was no reason to think the imperial system would come to an end.

# 1

# Foundations

Imperial Chinese government developed out of several, sometimes antagonistic, ideological positions. At its simplest, this can be reduced to the pairings of law versus morality, and centralization versus local power. Imperial governments naturally emphasized centralization over local power, and ruled by laws while claiming that their power stemmed from the moral attainment of the ruler. The possession of Heaven's Mandate, a Heaven-bestowed right to rule, was the source of a dynasty or emperor's legitimacy. Moral attainment received heavenly sanction, and loss of morality would induce Heaven to withdraw that mandate and bestow it on someone worthy. Law ultimately flowed from above, through the ruler and his officials to the common people.

The concept of Heaven's Mandate was created by the Zhou dynasty after it overthrew the Shang dynasty in 1046 BCE. Heaven's Mandate was therefore directly connected to success in battle or battles leading to the creation of a new ruling regime. Military success was proof of possession of Heaven's Mandate gained by moral attainment. Right made might, and might proved that one was right. Not surprisingly, this piece of political propaganda was so effective and useful that it was included in imperial ideology for every dynasty until 1912.

Political and military power were expressions of the fundamental morality of the ruler, who became Heaven's Son by holding the mandate. Ideologically, the power of the ruler, his court, and the state was all about morality rather than laws or the strength of his army. Before the beginning of imperial history in 221 BCE, the Zhou king, Heaven's Son, primarily asserted spiritual authority over the disparate states and fiefdoms of China. There was Zhou morality and its concomitant mandate, but there was no Zhou law spread uniformly across the lands under Zhou spiritual authority. Rather, individual states had their own laws within their borders.

All of that diversity among the states under the Zhou king was abolished at the beginning of the imperial era. When the Qin state conquered all of the other Chinese states in 221 BCE, unifying the Chinese ecumene, its ruler assumed a new title: *huangdi*, august thearch, or, as we have translated it in English, "emperor." The new ruler was not only the spiritual center of the Chinese universe but also the highest temporal authority, whose laws were now universally applied. The Qin dynasty sought to unify all practice, from axle width, to coinage, and on to writing, throughout its territory. The moral reach of Heaven's Son, the emperor, was now the same as his legal reach. Power of all kinds would now be centralized.

The Qin dynasty did not last very long, and was replaced by the Han dynasty, which ruled, with a brief interregnum, for four centuries. The Qin and Han were often contrasted, with the former's harsh Legalism and centralism set against the latter's benign, Ruist (Confucian) moral rule and localism. Although this characterization was a product of Han interpretation or propaganda, it retained a certain currency for the rest of imperial Chinese history. Its power and influence were due to the later cultural centrality of the historian Sima Qian (145/135–86 BCE), a man whose significance we will return to repeatedly, and his major work, *The Records of the Grand Historian* (*Shiji*). It was Sima Qian who established many of the categories of thought, and who presented Qin rule as singularly harsh. Recent archaeological finds, however, have shown Qin law to be both less harsh and more consistent with the other states than Sima Qian led us to believe. Just as critically, all imperial Chinese governments relied upon legal codes to manage their territories and populations.

In Sima Qian's telling, the Qin state was closely associated with Legalism, a system of thought that insisted upon the strict application of laws by the government. Order and productivity were the result of rewards and punishments. The Qin state ultimately conquered all of its rivals and unified China by a system that rewarded success in war and agricultural production, and harshly punished failure in battle and violation of law. Qin and Legalist harshness were thus portrayed as a necessary means to unify temporal authority, but an unsustainable method of rule.

The Qin also centralized authority, leading to discontent in the formerly diverse and locally ruled regions of China. The Han dynasty initially took a step back from Qin centralization, awarding large personal domains to generals who had founded the dynasty as well as imperial family members with their own lands. It would take the Han court decades to claw back control from those local power holders.

Although Qin law was retained, its enforcement was officially relaxed. At first, court ideology, to the extent that it had any, was Daoist. That changed in 135 BCE when Ruism was adopted as the main ideology for officials. Ruism provided a moral cloak for the Legalist functions of the state, and meshed with imperial ceremonial practice (Ruists had long functioned as ritual specialists for the court). Nevertheless, there was an inherent tension between the Ruist and Legalist concepts of how a state should function. That tension was now built into the basic Chinese imperial program.

The authority of Master Kong (Confucius) only grew over the two millennia of imperial Chinese rule. The record of his teachings, the *Analects*, would have been familiar to almost any literate person, and its positions on morality and ethics tremendously influential. Most government officials would have accepted Master Kong's views without question, or were at least capable of framing arguments in those terms well enough to pass the civil service exams and function in government, leaving aside whether a given official in fact lived up to those ideals.

In terms of the relative values of morality and law, Master Kong made two important points. First, he argued that rule by laws makes people avoid doing wrong only to avoid punishment, while rule by morality teaches people a sense of shame, which keeps them from doing wrong. Second, he said that while he was no better than anyone else when acting as a judge, the real goal was for there not to be any cases to adjudicate. These two points combined to cast the notion of rule by law in a negative light. There was nothing positive about ruling through laws, or even being an effective judge in applying those laws. The very act of applying laws was an indication of moral failure and misrule. Certainly, no Chinese emperor would ever wish to be given the sobriquet "law giver."

Yet even before the imperial era, Master Kong assumed that kingdoms would use laws to govern their subjects, and officials would serve as judges in applying those laws. Laws were not alien to Master Kong, he simply argued that a legal system would not make people better. A ruler's goal in the Ruist system was to insure the common people's livelihood and morally improve them. A morally superior people would also be easier to rule. The ruler served the people and thus served heaven. Instead of laws, a ruler should use rites (*li*) or ritual to order society. These practices, manners and ceremonies honored the worthy, diminished the unworthy, and transmitted correct values throughout society. The constant reinforcement of these norms through the ruler's behavior and that of the elites would not only transform society into a peaceful, prosperous community, it would also attract people from outside the state, eager to be ruled by a benevolent monarch.

As a rites-based society, the ideal Ruist state did not require and was indeed opposed to a rigid centralization of power. The ruler spread his influence through the charismatic power of his virtue, and the selection of like-minded officials. Those officials acted correctly in their local capacities without need for central direction; they had great autonomy to do what was right as they saw it. Even close court officials, chosen by the ruler for their moral attainment, were allowed great latitude in carrying out their duties. The ruler himself acted as a moral center while making few direct policy decisions.

In contrast to Ruists, several thinkers later grouped together as Legalists envisioned a centralized, rules or law-based state, which concentrated all power in the hands of the monarch. The function of the state apparatus was to carry out the ruler's will as he chose. Heaven had nothing to do with legitimizing the state or ruler. A monarch proved himself by imposing order on his government and people through strictly enforced laws. Officials were held to clearly delineated responsibilities and promoted or demoted according to whether they met those requirements. They were also required to not carry out tasks beyond their responsibilities. In one famous example, a ruler who fell asleep drunk woke to find himself covered with his cloak. However, the Keeper of the Hats had placed the cloak on him because the Keeper of the Cloaks had failed to do so. The ruler punished both men, one for failing in his duty and the other for overstepping his responsibilities.

Centralized control was paramount in the Legalist system. All power and authority stemmed from and returned to the monarch; officials had no real autonomy, only specific tasks to carry out. Order was created by the universal and strict application of laws, rewards, and punishments. With the exception of the ruler, everyone was subject to the laws without regard to family, relationship, or status. The state established what was right or wrong. In the case of the Qin state, whose political culture developed during a period of continuous interstate warfare, success in war and agricultural productivity were paramount.

Master Kong provides us with the best contrast between the Ruist and Legalist perspectives on society. The *Analects* contains a story in which a man boasts to Master Kong that in his state the people are so upstanding that if a man committed a crime, his son would report him to the authorities. Master Kong was unimpressed, preferring his own state where a son would protect his parent. Imperial governments never resolved this tension between the impartial needs of the state and its relationship to individuals, and the ideal familial relations that made for a stable, moral society. Was a subject an individual whose key responsibilities were to the state, or were they part of a family unit with collective responsibility to the state? Was one's father more important than one's ruler? States simultaneously saw the family as the basis of all morality and order while also demanding individual loyalty and obedience.

The conflict between ruling the individual and the family was another reflection of the central versus local authority issue. Just as an enfeoffed lord or local strongman mediated the relationship between the individual and the central ruler, so too did the family. The Qin creation of imperial China established an unmediated relationship between emperor or state and individual throughout the empire. Law governed everyone equally, at least in theory, but social and moral practice insisted on the primacy of the family and the authority of one's father. Therefore, the law had to support parental authority as a matter of fundamental morality and maintain a state role within the family.

## The Imperial State

Chinese empires were bureaucratic institutions from the very beginning. The compromise between law and morality created a state apparatus run by regulations and the application of a legal code mitigated by the judgment of the officials who ran the government. Thus, during the Han dynasty, the government could make Ruism its official ideology without changing the underlying system of regulations and laws. Obviously venal or immoral officials were sometimes removed, but honest and decent officials were unexpected enough to warrant comment. By late imperial history, officials were almost universally expected to be corrupt. If this is less clear for earlier periods it is likely a reflection of our limited sources or different concepts of corruption. Some level of what a twenty-first-century person would characterize as corruption would have simply been the way government was expected to function. Bribery and bias were the norm. The wealthy and powerful were supposed to get preferential treatment. The only issue was at what point did corruption become intolerable to the government.

Morality was important in the construction of state power. A government official was not just expected to withdraw from government to mourn a parent for three years (less for a mother in early imperial history), he was morally required to do so. An emperor could declare an official so critically important that this moral requirement could be set aside, but this was functionally confined to generals on campaign or in the midst of a military crisis, and seldom invoked. Even the suggestion that an official might ask for or accept an imperial indulgence regarding mourning a parent would be enough morally to disqualify him from government service. Officials were supposed to be moral men. Although most officials recognized that Ruist, educated men frequently acted in an immoral fashion, they persisted in maintaining the ideal that this was unacceptable.

Ruist morality was a basic component of literate education, but legal training was not. Most people were likely aware of common penal laws, such as prohibitions against murder or stealing and their punishments, but civil laws for things such as inheritance and property disputes could be considerably more obscure. There were also highly specialized administrative laws of direct concern to officials. To further complicate matters, local magistrates were not only responsible for adjudicating cases, they were also responsible for investigating them. The application of laws was, therefore, often in the hands of a poorly trained, though highly educated, man whose political career depended upon maintaining social order as much as administering justice. Earlier in imperial history, magistrates were simply responsible for fewer people and faced fewer complicated cases. Yet all cases were subject to possible review by higher levels of government. This was more likely if the plaintiffs involved were themselves educated or powerful people who knew how the system operated and could draft appeals—or hire someone to do so. In later imperial Chinese history, the larger populations a magistrate had to govern created a sub-class of specialized secretaries knowledgeable about individual areas of responsibility. They advised the magistrate on the law and best practice with regard to rulings.

Sentences involving capital punishment were always carefully reviewed by the central government because executing innocent people might arouse heavenly displeasure. On rare occasions, a few generals were given special authority for the duration of a campaign to execute first and inform the throne afterward. The fact that for generals on campaign, who were otherwise authorized to have their soldiers kill whomever military exigency required, it was important not to apply military law carelessly, shows how seriously capital punishment was taken. While punishments were often brutal—executions, canings, and so on—they were applied carefully and in accordance with the laws.

The earliest complete extant legal code is the Tang dynasty code promulgated in 624 CE (a partial collection of Qin law and case books was recovered from a tomb in 1975). It derived from earlier codes, and shows the deep influence of Ruist morality, the extant classics of which were, of course, far older. The Tang Code organized 500 sections into 12 volumes, and no longer contained punishments like castration or physical mutilation. Some applications of caning or flogging might have been severe enough to result in permanent physical damage or death, but someone sentenced to be caned was not supposed to die, since they had not been sentenced to do so. Such an outcome might initiate an investigation and punishment for the official involved. The laws were supposed to work as they were designed. As the removal of some earlier punishments suggests, once judgments about acceptable kinds of punishments shifted, the laws were modified fully to instantiate those changes.

The vast size of imperial legal codes and the limited availability of printed copies until well into the thirteenth century, at the earliest, calls into question how carefully the full range of those laws could be applied. At the same time, the need to interpret those laws correctly and apply acceptable judgments gave rise to collections of case law. The first book in this tradition was *Enlightened Judgments* (*Qingming Shupan Ji*), published in the thirteenth century. As with the legal codes themselves, most heavily based upon their predecessors, *Enlightened Judgments* was as much a reflection of morality, and identifiably Ruist morality, as the needs of imperial government. A very practical manual, it contained concrete examples of the sorts of cases a magistrate might actually be called upon to adjudicate and what constituted an ideal, or at least acceptable, ruling.

Official regulations, the rules that structured the government, were just as important for the functioning of society as the penal and civil codes. Regulations defined the relationship of the central government to local governments; the autonomy of powerful local authorities, including landlords, temples, and monasteries; the duties of officials at all levels; and economic relationships throughout the empire. For example, in early imperial China most of the adult male farming population owed the government annual labor in the form of military service or construction work. Military service became less common during the Tang dynasty with labor service gradually dropping out in favor of taxes shortly afterward. In the second half of imperial history, most military service was either professional or hereditary—the sons of officers usually became officers, and the sons of common soldiers became soldiers. Similarly, starting in the Tang, government officials began to be selected through an exam system, creating the ideal, if not often the practice, of great social mobility. Regulations promoted the ideal of a meritocracy.

The connection between Heaven, the emperor and the law was particularly clear in the Great Ming Code, first promulgated in 1367 and finalized in 1397. It intensified the centralization of all-encompassing spiritual and temporal authority in the hands of the emperor and his government. The Great Ming Code explicitly tied its statutes to heavenly principle, and regulated social, moral, civil, and penal law. Its goal was to reform the empire at every level, not just enforce imperial authority, as reflected in its three divisions: Rituals for spirit world communication, ideals for humans and human society, and rules to improve the elite who stood between the people and the divine. A unified system from the spirit world down to the common people, Ming law saw the emperor and his officials (including the elite) administer heavenly principles in human society. For the Ming government at least, it removed any tension between morality and law. As a consequence of its comprehensive renewal of earlier traditions, Qing dynasty law closely followed the Great Ming Code, as did other East Asian states.

## Law and Morality in Reality

While the Great Ming Code may well have resolved the conflict between law and morality for the government, the common assumption among the populace as portrayed in fiction (which may well have actually been the opinion of elite writers), was that most officials were corrupt and that the law served the powerful at the expense of the weak. Many literati also felt that government service was inherently corrupt. Government service was the highest goal of Ruist literati ideology, but it was also understood from very early times to be politically and morally fraught. Even before the beginning of the imperial era good men at court despaired of corrupt politics.

Other educated men throughout imperial history also withdrew from government or chose not to enter it because they could not reconcile morality with the necessities of officialdom. "The Seven Sages of the Bamboo Grove," a group of seven men in the third century who chose reclusion, drinking, and conversation over government service, exemplified the attractions and criticisms of withdrawal. For many literati their example was an attractive idea, separating oneself from responsibility and the very real dangers of court life. Other literati criticized the Seven Sages for not using their talents to improve a failing age. Similar criticism would be lodged against other educated men in tumultuous times. Consequently, for many of those men, including at least one of the Seven Sages, Ji Kang, the attempt at withdrawal actually led to execution. Rulers did not like or accept a literatus refusing to serve on moral grounds. Leaving the court required far more bravery than some later critics were willing to credit.

The literati, as a group, asserted their own primacy over morality eight or nine centuries later, during the Song dynasty. This decentralization of moral authority took place under a highly centralized government or, perhaps, in reaction to that centralization of temporal authority. For Song literati it was they and not the emperor who decided moral issues. This assertion became fundamental to literati identity from then on, placing the subsequent Great Ming Code's insistence on imperial control over everything spiritual, moral, and legal into perspective. Moreover, during the Song the literati as a class expanded beyond the limits of official families (those with family members who had served in government) to include educated men or even locally powerful families whose lifestyles demonstrated literati values. Ruist literati or gentry pursued, so they claimed, moral attainment, defined by learning, manners, and proper social engagement. These gentry, an extremely hard to define group, effectively ruled locally by their power and influence, making pretensions to morality and legality in their actions. They carried out charitable works, promoted education, organized local defense forces, and led society by their Ruist inspired and sanctioned behavior, and believed that this justified their power.

## Conclusion

There was a continual struggle for moral and legal authority in imperial China. The court tried to centralize this authority in the person of the emperor and to regulate the empire through government power. Countervailing forces within the government and outside it tried to decentralize those same authorities, and to influence society and culture in ways if not contrary to the emperor then at least independent of him. Because the emperor and the dynasty claimed legitimacy through possession of Heaven's Mandate there was also a spiritual or religious aspect to their claims of authority. The Great Ming Code even regulated Buddhist and Daoist practice, popular religious practice, and prohibited perverse religions. While earlier dynasties had no compunctions about their right to regulate religions, the effects of their efforts were often limited.

Law and morality were not simply tools to control the populace and maintain state power; many of the men in charge genuinely sought to create a better society. Some were cynical, but most were not. If exemplary moral attainment did not immediately and of itself transform society then law as an expedient was necessary and acceptable. Master Kong himself did not reject the application of law and punishment. He did not even claim that he had succeeded in eliminating the need for laws. A moral society was as much a goal as a well-fed and well-mannered population. In practical terms this might have been unattainable, but a good ruler and a good government tried hard to achieve it nonetheless.

As imperial power and centralization grew over the course of imperial history, the literati and the larger gentry class carved out their own spheres of influence and power. These were largely local and, as long as they did not link up to challenge imperial control, the central court ignored them. The average citizen was far more subject to these local authorities than to the imperial court or even the government. Claims to morality and law persisted at all levels of power without necessarily holding much actual sway. A wealthy landlord who educated his sons and acted like a gentleman did so not because he needed to bolster his moral influence with his tenants, but in order to accommodate himself to the government's claims of proper behavior. By going along with the government's pretensions, he could wield real power. Yet, from the government's perspective, convincing local elites to at least appear to conform to its moral, legal, religious, and cultural demands demonstrated effective political power.

At root, the political power of dynasties was based upon the unprovable claim that Heaven had somehow given the emperor or his family the authority to rule. This worked as long as the many powerful people at all levels of society cooperated enough to support the government's power. At the lowest level were magistrates, appointed by the central government and sent out to counties all over the empire—except where they were from, to prevent favoritism. Above them, interlocking administrative structures provided supervision through circuit intendants or inspectors for things like taxes, education, or military affairs, ensuring no single official held too much power. This was all then subordinated to the central government bureaus in the capital under the direct control of high officials who reported to the emperor himself. Law and its enforcement was critical for the government, but claims to morality were more important for the ruling class, particularly the imperial family. Governments might run on law, but dynasties survived on morality.

# 2
# Dynasties

Imperial dynasties were vague, abstract focuses of loyalty, historical categories, and institutional names for the familial descent of their rulers. They had territorial aspects (discussed in Chapter 3) and political standing. Their general form was fixed early on by the Qin dynasty, with tremendous diversity in their specific structures varying not just between dynasties but over the course of dynasties as well. As the focus of elite service, dynasties sometimes competed with particular rulers for loyalty. In their own way, dynasties were as ambiguous and contested as the borders of the territory they claimed.

Every dynasty claimed possession of Heaven's Mandate, creating an inherent historical claim to orthodoxy that could be lost just as it was gained. In order to gain Heaven's Mandate due to a founder's moral attainments someone else had to lose the mandate due to moral failings. In principle, the idea of one dynasty losing the mandate and another picking it up was simple. Historical reality was often more complex, however, with messy periods of transition between large, stable governments, or multiple governments all claiming to possess the mandate.

Although we often describe China as regularly maintaining stable governments ruling over most of the Chinese ecumene, there were in fact only a half-dozen dynasties that could claim to have done so. The Qin dynasty (221–206 BCE) that began the imperial age was quite brief, but gave way to the Han dynasty (206 BCE–220 CE), which ruled, with one significant break, for four centuries. The Sui dynasty (581–618) brought the instability of the Northern and Southern dynasties period to an end, but quickly fell, having set the stage for the Tang dynasty (618–907). A half-century of disunity followed the Tang fall, until the Song dynasty (960–1279) recreated much of their empire. The Mongol Yuan dynasty (1279–1368) was itself fairly brief, but was also part of the greatest empire in the history of the world. The Mongol empire (*ülüs*) stretched across Eurasia, encompassing a wide variety of peoples and polities all tied to the central court of the Mongol Khagan/Qagan. It was followed by the Ming dynasty (1368–1644) and, finally, the Qing dynasty (1644–1912).

Even this description of "major" dynasties is vastly oversimplified, leading many to throw up their hands at the overwhelming list of governments and territories of imperial history. It often seems as if experts casually mention different dynasties with the expectation that everyone knows when they ruled and how they have been characterized. Partly this is due to making generalizations about China, Chinese culture, and imperial China, as if its long history were an unchanging, consistent mass, and that there was a classical culture which proceeded with only minor differences up to the present day. But Chinese history is structurally different from that of the West insofar as a half-dozen governments controlled large portions of the Chinese ecumene over some two millennia, masking regional differences and making empire a normative ideal. That political level of unity seldom existed in Europe after the fall of Rome, though the ideal remained, and even the Catholic Church's spiritual authority eventually fragmented.

Politics in that sense dominates the idea of the imperial dynasty. A dynasty was a political entity that drew upon cultural norms to legitimize its rule. Those norms included not just imperial rites, but also claims to Buddhist and Daoist support, and Ruist social and political ideas. The imperial rites often appeared to be Ruist in origin because they did not invoke Buddhist or Daoist beliefs, and because of the extensive involvement of Ruist scholars as ritual specialists (indeed, the Ru were originally known as ritual specialists). These rites were an extension of a dynasty's claims to Heaven's Mandate and the ruler as chief practitioner. Ordinary people did not sacrifice to Heaven or the Marchmounts, the mountains that marked the ritual boundaries of the empire, but Heaven's Son did. Only he could perform the Feng and Shan sacrifices, the highest imperial rites, before Heaven and Earth at Mount Tai.

Dynasties were, and are, also important as historiographical categories. Imperial governments produced enormous amounts of official documents that were used in subsequent dynasties to write histories. Dynasties therefore also formed natural historiographical groupings, with documents, historians, and personnel linked through proximity and politics. As a consequence of this, modern historians often identify themselves as historians of a particular dynasty, and this is reified in scholarly publications that focus on a single dynasty.

## Similarities

Despite the distinct histories, institutions, and cultures of the many imperial dynasties, there were also a number of fundamental similarities that make it possible to discuss them together. This form of rulership is also what allows historians to claim dynasties ruled by non-Chinese rulers are part of Chinese history. Since those rulers explicitly created Chinese-style dynasties claiming to hold Heaven's Mandate and were led by someone claiming to be Heaven's Son, their dynasties and governments were "Chinese" even if they identified themselves otherwise. The imperial family of the Qing dynasty, for example, were clear that they were Manchus, but Chinese officials serving under them saw the dynasty itself as Chinese.

All dynasties constructed bureaucracies to administer their empires while also retaining a system of noble titles. In some dynasties those titles were attached to actual grants of land with real power over those lands, as well as the income derived from them. Most, however, were markers of status that fit within a complex bureaucratic system. Imperial Chinese government, like imperial Chinese society, was extremely formal and hierarchical, particularly among the elites. Some titles would not only carry forward to descendants but also backward to retroactively, and even posthumously, elevate a man's parents in rank. The parents of a woman who became an imperial consort, married into the imperial family, or became empress or empress dowager might also receive titles.

The bureaucracy itself was one of the most distinctive and consistent markers of Chinese empire. Like all bureaucracies, Chinese dynastic governments generated an enormous amount of paperwork (though in the case of some of the earlier dynasties like the Han, the "paperwork" was written on wooden slips). This paperwork was part of the centralizing tendency of imperial governments; reports were collected and flowed up the hierarchy to the center, and orders flowed from the center out and down to subordinate offices.

The highly literate system of imperial governing required concomitantly well-educated officials to compose and digest the flow of reports and orders. These officials learned the designated spoken language of the court (the Mandarin of each dynasty was different) as well as the proper written forms of Literary Chinese for government service. The issue of translation between officials and local populations was barely mentioned. Officials sent from elsewhere in the empire could not be expected to speak the local language and likely relied upon the permanent local staff for translation. Local educated elites would also have spoken that dynasty's Mandarin, making them not only socially and culturally similar to an official, but the only people, beyond the local staff, with whom an official could speak directly. Officials therefore tended to sympathize with local elites who shared their values and background.

Imperial Chinese governments were exclusively staffed by men, and every emperor was a man, with the exception of Wu Zetian at the turn of the seventh century. The inner palace staff, that is the servants within the imperial living quarters, were usually eunuchs, with large groups of female servants and other ladies in waiting. There were also female-staffed bureaucracies within the inner palace that attended to the extensive administrative requirements of the imperial household. Unfortunately, we know very little about these bureaucracies, or the balance between eunuchs and female inner palace officials. Their main concern was ensuring the patrimony of any children born to either the empress or the other imperial consorts. There were theoretical ritual limitations to the number of consorts an emperor could have, but no emperor even came close to those limits.

Genealogy was critical to the maintenance of a dynasty, but the understanding of who was qualified by birth to succeed to the throne could be quite broad if circumstances required. All male children of a Chinese emperor had an equal claim to succeed their father. There was no rule that required or preferred the offspring of the empress; the male child of any imperial consort could become emperor. In several cases the male offspring of an emperor's brother, though never sister, could succeed to the throne if an emperor lacked a male heir. It was even possible for an emperor's brother to succeed to the throne as Heaven's Mandate established a line of descent from a family rather than an individual.

In practice, the broad pool of possible imperial successors could produce ruinous power struggles. Emperors were reluctant to limit those struggles by openly designating their heir because

## WU ZETIAN

Wu Zetian (624–705) was the only woman in Chinese history to become emperor. She became an imperial consort of the second Tang emperor, Taizong, at the age of fourteen. When he died in 649, she managed to avoid permanent retirement to a Buddhist convent and returned to the palace as the imperial consort of his son, Emperor Gaozong. This was highly irregular, suggesting an unfilial and functionally incestuous relationship with a father and son. She gradually overcame her rivals among the emperor's consorts, as well as the sitting empress, to become empress herself in 655. Her power continued to grow as she became directly involved in imperial decision-making, with dozens of high-ranking officials and members of the imperial family murdered or executed during her rise.

When Gaozong died in 683, Wu Zetian became regent for her son, Emperor Zhongzong, who she soon deposed and replaced with a younger son. In 690, the younger son abdicated the throne to her, whereupon she established a new dynasty, the Zhou dynasty, with herself as emperor. In 705, when she fell ill, a coup overthrew her and restored Emperor Zhongzong to power. She died later that year, and was buried as empress with her husband Emperor Gaozong.

that would create an alternative focus of loyalty. Of course, hiding who the designated heir was, or entrusting the information to selected officials, left open the possibility of secretly substituting one family member for another. No dynasty ever completely solved this problem. Nor was it clear what to do if an emperor became incapable of ruling. When Song emperor Yingzong (r. 1063–67) fell ill soon after taking the throne, Dowager Empress Cao took over as regent. Despite the fact that she had ruled well, she had to be forced from power when the emperor recovered. It was simply unacceptable to the Ruist officials that Cao continue in control once the emperor was well, even if he were less competent than she. Some emperors went through periods of regencies, but there was no standard age for ending a regency and very few dowager empresses or other regents were quick to relinquish power. Certain emperors had to seize power from their own regents by force.

Whatever the symbolic supremacy of the emperor, the practical aspects of his power were always subject to gaining and retaining the support of powerful figures and, unsurprisingly, actual control over the person of the emperor was key to wielding power. An emperor who was essentially unfree, because he was physically under someone's control, was merely a figurehead. In theory his power was absolute, but in practice exercising that power required considerable political skill.

## Differences

The consistent broad outlines of dynasties masked significant differences between these political units. Dynasties varied widely in the actual institutional structures of their bureaucracies, the organization of their militaries (see Chapter 4), their systems for recruiting officials, and their legal systems, to name just a few areas. The great Han historian Sima Qian included a section on institution in *The Records of the Grand Historian (Shiji)* because the evolution and function of Chinese institutions was fundamental to understanding the operations of government. Indeed, administrative guidebooks were important for officials trying to understand how the dynasty they worked for functioned.

The administrative structure of a dynasty was a combination of inherited institutions and new priorities. No dynasty created its institutions from scratch after its founding, but each made significant changes in response to the concerns of the founders. Many also made important changes in their institutions over the course of their existence. At the same time, a dynasty's later emperors and officials usually argued that they were not, in fact, changing the institutions set up by the dynastic founder. "Innovation" was not a positive concept for dynasties because Ruist ideology was fundamentally conservative.

The Qin dynasty was unique among imperial governments in highlighting its innovations and attacking traditionalism. It sought and achieved a much greater degree of centralization than its successor, the Han. Yet as much as Liu Bang, the Han founder, was forced to enfeoff many of the generals who supported him in the struggle to establish the dynasty, as well as many of his own family members, that was not his preference. Circumstances forced Liu Bang to move away from Qin centralization because it was ideologically expeditious to do so. Many people rejected the way the short-lived Qin dynasty had been run and sought a return to the pre-imperial, decentralized, or "feudal," system. Liu Bang, in contrast, while making the necessary concessions to establish his dynasty, spent most of his reign putting down rebellions of enfeoffed power holders. That pattern culminated in the Rebellion of the Seven Kingdoms in 154 BCE, when Emperor Jing (r. 157–141) defeated these imperial princes and brought Han centralization closer to that of the Qin. Fiefs still existed, but their power was greatly diminished.

Later dynasties also varied in their degree of centralization. Magistrates in counties had great local power, but the manner in which the higher levels of government were organized changed in response to growing population and shifting threats. Supervision above the level of the county or prefecture could be vested in intendancies, areas under the supervision of an official responsible to the central government called an "intendant," for various matters, or provinces. In the Song dynasty, there were four different intendancies for the central government's military, fiscal, judicial, and supply matters. These intendancies did not coincide with one another, and were staffed by separate officials. This balanced the need for supervision of the lower levels of the bureaucracy against the fear of any official gaining control of sufficient resources to threaten the throne.

The Mongol Yuan dynasty, a Chinese-style imperial government institution created by Khubilai Khan to help rule over his Chinese subjects, was organized quite differently, with its power structured more as an occupying force. Many ad hoc offices were controlled by Mongols or Central Eurasians who held their positions because of their ethnicity and loyalty to the Mongol *ülüs*. Government at the local level was a holdover from what had existed before the Mongol conquest, and at the highest level was centered on the imperial household rather than a bureaucratic set of government institutions. Apart from establishing the ritual aspects of a Chinese dynasty, the Mongols governed very differently than their predecessors, the Song.

The Ming dynasty, in turn, shifted back to a more purely bureaucratic system, but eliminated the position of chief minister to prevent a powerful official from usurping imperial prerogatives. Chief ministers or prime ministers led the government's bureaucracy, channeling information up to the emperor and orders down to the officials. This critical position allowed them to control what an emperor knew and interpret how his orders should be carried out. Abolishing the position of chief minister meant the Ming emperor had to deal directly with various departments of the central government, creating an overwhelming burden of paperwork. Even though a group of secretaries was established to help him, the Ming system still required the emperor to play an active role, and when some failed to govern, the bureaucracy ground to a halt.

The final imperial regime, the Qing, established when the Manchus conquered China, combined some of the aspects of an occupying force, like that of the Mongols, with a centralized Chinese bureaucracy. Manchus were the most trusted officials in almost every position, often serving alongside Chinese counterparts. They were all part of the eight Manchu banners, the groups, marked by colored flags or banners, into which Manchu society was organized. Individual banners served as social groups that mustered troops when required by the ruler, and in turn received lands and goods for, in theory, the banner as a whole. This privileged group formed the ethnic core of the dynasty even as individual Manchus lost their martial traditions as the dynasty progressed. So much larger was the population of the Qing empire, compared to earlier dynasties, that additional layers of government were added between the local levels and the central government. Regional officials overseeing provinces or the collection of taxes for two or more provinces were responsible for millions of people.

## CIXI

Tsysi taiheo, a Manchu of the Yehe Nara clan, known more popularly as the Empress Dowager Cixi, was born November 29, 1835. She became an imperial consort of the Xianfeng emperor, and gave birth to the future Tongzhi emperor in 1856. Once her son became emperor at the age of five, the now empress dowager joined Empress Dowager Ci'an (the previous emperor's empress) and Prince Gong (the previous emperor's sixth brother) in a regency. The Tongzhi emperor died childless in January of 1875, less than two years after taking power, and so another regency headed by the two dowager empresses installed the Tongzhi emperor's three-year-old cousin, the Guangxu emperor, as the new ruler. Ci'an died soon after, leaving Cixi in control.

As dowager empress, Cixi presided over the collapse of the Qing dynasty. Her dominance of the government was based upon a careful balancing of political forces that made it difficult to respond to a series of crises stemming from internal and external forces. While she does not deserve all of the blame she has received for the problems of the late Qing, it is also true that she failed to respond effectively to the problems besetting China at that time. She died in 1908, having likely ordered the poisoning of the Guangxu emperor the day before.

# The Other Dynasties

Aside from the half-dozen "great" dynasties of the imperial period, there were also dozens of other polities that controlled often quite large pieces of territory for briefer periods of time. Those states and kingdoms reflected the similarities and differences in organization, ritual practice, and culture of all imperial Chinese polities, large and small. Because they were shorter-lived, or only controlled parts of China, they have always received less scholarly attention, even before the twentieth century. They could neither be held up as paradigmatic examples of a unified, long-lasting, territorially "complete" empire, nor as complete military and political failures. And because those lesser polities were difficult to fit into the orthodox succession of dynasties passing Heaven's Mandate from one ruler or ruling family to another, they also had a more complex historiography.

However, neglecting those other dynasties tends, intentionally or not, to reify two critical, related fallacies. First, that imperial China had an unchanging culture separated by episodic changes of ruling family. Second, that China was always a naturally unified political and territorial unit. By skipping over the times when this was not the case, it is easy to assert that it was never the case. "Conventional" and "orthodox" periods of imperial rule were separated by "transitional periods," "times of chaos," and "interregnums," as if these disunified periods were anomalous. Yet China was frequently ruled by regional powers or unstable regimes.

China was not a natural political unit that could be assembled without great effort. In some periods, it was simply impossible to unify all the areas that were home to people who would have identified as Chinese. Moreover, north China was closely linked with steppe groups whose waxing or waning political cohesion directly affected their influence in Chinese affairs far more than anything the Chinese did. It was much more difficult for steppe-focused regimes to extend their power south of the Yellow River because that required both a navy and substantial infantry. The disparate modes of warfare that were necessary to conquer all of China favored fragmentation rather than unity. Cavalry was critical to the conquest of the north; naval vessels were required for spanning the great rivers running east–west, and campaigning in the south; and infantry were needed to fight in the south, and to capture and hold cities. Regional powers tended to be very accomplished at the mode of warfare most effective in their area, meaning they were strong locally but less able to project power beyond their base.

A deeper problem from the cultural perspective was what in fact defined China or Chinese culture during periods of fragmentation. If multiple political centers could all claim to be Chinese, did that mean that Chinese culture did not overlap precisely with political power? Imperial ideology claimed centralized control over cultural orthodoxy. Even pre-imperial political ideology asserted the ritual, spiritual, and cultural centrality of the Zhou king, and before him the rulers of the Xia and Shang dynasties. A regional warlord might not claim to be Heaven's Son or possess Heaven's Mandate, but he could still challenge imperial ideology if he did not recognize the claims of someone who did. And what if there were multiple claimants to be Heaven's Son?

Orthodox ideology asserted that Heaven could have only one Son, and that the Mandate could only be held by one family. Periods of disunion forced statesmen to compromise on rituals, or create polite fictions to maintain the dignity of multiple rulers. After a military face-off at Chanyuan in 1004, where the Song and Liao emperors confronted each other with their armies before negotiating a settlement, the Song dynasty was forced to accept the imperial dignity of the Kitan Liao dynasty. The two courts agreed upon the fiction that the imperial families were related, despite one being Chinese and the other Kitan. (The Kitan were a proto-Mongolian group who originated in the steppe, and became more politically coherent during the late Tang dynasty.) Although galling to Song statesmen, it maintained ritual propriety and more than a century of peace.

The two main periods of disunion during imperial history were the Six Dynasties period (222–589), and the Five Dynasties and Ten Kingdoms period (907–60). Confusingly, the Six Dynasties period can also be divided into the Three Kingdoms (222–80), Jin dynasty (265–420), the Northern and Southern dynasties (420–589), some of which overlapped with Sixteen Kingdoms (304–439). This confusion makes it impossible to craft a clear and comprehensive narrative of the period. Similarly, the Five Dynasties and Ten Kingdoms period covers the five regimes that ruled north China, which the Song dynasty grew out of, as well as the unstable mix of southern and western regimes, states, and kingdoms that the Song eventually conquered. To add to that confusion, the Kitan Liao dynasty, while mostly in the steppe, was also involved with north China. These were important eras for high culture, though historically complicated.

Periods of disunion highlight the regionalism and separations that have always existed across the Chinese ecumene. Dynasties intentionally downplayed the natural fissures in Chinese society for political reasons. By insisting that political unity was based upon cultural and social unity, they formed the necessary institutional superstructure to govern a vast empire. The persistence of the possibility of disunion, let alone its actual appearance, undercut the basic dynastic premise. The great dynasties did not see themselves as empires occupying and controlling disparate peoples and places. The other dynasties, however, along with minor states and kingdoms, were bare spots in the tapestry of orthodox imperial ideology.

# Conclusion

Dynasties were political units that claimed a consequent and inherent territorial, cultural, and historical unity. The ruler of that unified polity had, in theory, complete control over laws, culture, and society. At the same time, the ruler claimed continuity with earlier culture. Like any successful political institution, Chinese dynasties were both dynamic and conservative. None lasted forever because none were able to adapt continually their real institutions and personnel to changing circumstances. However, larger concepts persisted for two millennia because they were clearly useful and functional. Dynasties provided a focus for political loyalty that allowed some flexibility in choosing leaders, while also limiting instability by narrowing the pool of possible rulers. In their simplest form, the dynastic system of government provided a way to choose a unifying leader.

Power radiated out from the emperor's central court in the form of government institutions run by his chosen representatives. These institutions of power were instantiated through ceremonies and bureaucracies that cloaked the direct use of violence, internally and externally, and the state's coercive functions. All aspects of power—cultural, social, economic, and military—were vested in the state with the emperor as the final arbiter.

The solution to the problem of choosing a ruler followed the usual human pattern of relying upon descent from a single family. Chinese dynasties were no more effective in resolving the problems of blood descent than anywhere else in the world. No family, no matter how broadly defined, was ever able to supply an unbroken line of competent rulers. Functionally, there was no reliable institutional way to guarantee the accession of a competent emperor. The system only worked over the long term because the most important function of an emperor was to be legitimate. A rightfully chosen emperor, no matter how inept or dissolute, could maintain imperial and dynastic validity where a wrongfully chosen emperor could not.

The legitimacy of a competent founder remained with his descendants even as they failed to rule well. Large bureaucracies were often able to maintain functional rule even when a particular emperor proved inept because, excepting moments of real crisis, the truly critical role of the emperor was simply to sit on the throne. An actively bad ruler could wreak havoc with the state, but a merely mediocre ruler, as most were, sufficed in ordinary circumstances. Institutions served as the main prop to dynastic survival, insulating the operations of state from capricious emperors. Often the hope was that the idiosyncratic successes of an extraordinary founder and his founding officials could be transferred to the orthodox consistency of a mass of mediocre bureaucrats. The dynasties that could establish rule by bureaucracy lasted for centuries; those that did not fell soon after their extraordinary founders died.

Dynasties were reflections of their founders, and transferred personal loyalty to institutional loyalty. In modern terms, subjects could be loyal to the Chinese nation, but no such nation existed during imperial times. For most subjects, loyalty was not much of an issue as the government authorities were of marginal importance to their everyday life. The elites who studied to become government officials and responded to dynastically directed cultural and political changes were far more invested in the issue of dynastic loyalty. Officials were not supposed to serve another dynasty if their own fell, but could serve another emperor within the same dynasty. Dynasties thus simultaneously empowered emperors by concentrating power in their hands, while also rendering any particular ruler a transient placeholder.

As part of imperial Chinese history, dynasties exert their most significant analytic force by fixing historical categories. They are lines that surround sets of documents and groups of people, or provide the opportunity to cross those boundaries to show change or to make comparisons. As a set of imperial dynasties, the polities that make up imperial Chinese history assert both a break with pre- and post-imperial history, and continuity between those dynasties. The existence of China as a unit of analysis relies upon dynasties functioning as "Chinese" political organizations. When non-Chinese peoples established Chinese institutions of administration and ritual following their conquest of all or parts of the Chinese ecumene, they joined Chinese history and asserted the dominance of that system of rulership. Those non-Chinese people had very different perspectives on their own histories in China. Outside the realm of imperial Chinese history, or the perspective of Chinese culture, dynasties and all that went with them were far less important. These named ruling houses allowed imperial Chinese history to subsume a wide variety of peoples, governments, cultures, and territories under the unifying category of what in modern times is known as China. Yet, as we shall see in the next chapter, the territory of what we think of as imperial China was as varied as its dynastic structures.

# 3

# Borders

Imperial governments both rejected and imposed borders on their territories. They rejected the notion that their power should not extend over everything while struggling to face the practical limits of their authority. Without carefully marked and defended borders, people of different cultures could and did freely move in and out of imperial territory, however hazily defined. Internal borders, much better marked, separated administrative boundaries and areas of responsibility. Imperial governments vacillated between strong, effective regional controls, which might threaten the central government, and weak, less effective regional and local power, with overlapping and contradictory jurisdictions. Of course, the borders of the succession of Chinese dynasties varied considerably. Song dynasty China, for example never controlled the area around modern Beijing, or extended much into the steppe.

Borders were part of the terrain and environment. Like political and administrative borders in China, environmental borders were usually zones rather than lines. The territory of the core of China varies widely from north to south and east to west. As the territorial reach of different dynasties extended out from that core, their political authority encompassed even more varied environments, peoples, and cultures. Local identification was as important as cultural and ethnic identification, though most people in China never traveled far enough from their birth-places to even consider these identifiers as choices or distinctions. However, there were several kinds of people who regularly traveled across internal, and sometimes external, borders: officials, merchants, soldiers, and religious professionals.

Natural geographic boundaries were external to politically established borders. Unfortunately for the political authorities, these weren't often situated in strategically ideal places. There was no easily defensible line separating China from the northern steppe, for example, or major geographic features that clearly marked the edge of the Chinese ecumene. Steppe groups could, and often did, raid, invade, or otherwise intrude upon China's Central Plains unless opposed by significant military forces. Some of the states that did not even rule all of China could still be quite large and encompass diverse environments. But where states might be politically delineated by geographic features like mountains or rivers, economically and culturally coherent regions responded to those physical realities differently. Rivers, for example, were more likely to form the core of economically and socially integrated regions than boundaries between regions.

Geography was not, in this sense, always political destiny, even if it was usually economic destiny. Chinese armies and navies overcame all of the geographic difficulties of the Chinese ecumene to create unified, central rule, proving that even where there were natural geographic boundaries, they were no guarantee of security. Sichuan, which was surrounded by mountains, and only accessible by sailing up the Yangzi River through the gorges, or from the north over the trestle roads built into the sides of mountains, could form a separate state when there were weak powers in the rest of China. But when a strong power developed Sichuan was repeatedly conquered, as happened when the rising Song dynasty captured it in just over two months in 965.

Borders served two main functions, as instruments of control, and tools of imagination, with considerable overlap between them. The drawing and redrawing of borders for internal territories, as Ruth Mostern has shown for the Song dynasty, was a direct application of administrative oversight. Troubled districts requiring greater government attention had smaller units of administration to facilitate tighter supervision. Such supervision came with the threat of direct government force, which would ideally convince the local population to fall back into line. Exerting this level of control was more expensive, however, and often raised costs above revenues. On the other hand, tax returns could be maximized in peaceful localities loosely supervised in larger administrative districts, albeit at the cost of less stringent controls. All dynasties and states in imperial China shifted their internal borders between expensive control and greater tax receipts as local circumstances required.

At the same time, regions and districts retained separate identities based upon their histories, ethnic makeups, or outside connections. They often looked back to older place names, particularly as a way to, at least intellectually, separate themselves from the central government. Similarly, local gazetteers reinforced place identity by grouping together sets of people, sites, and events related to a named location. Like any border, what was not included in a local gazetteer was as important to delineating a place as what was included. A line then separated one place from another place, even if that line was only a written description based upon a historical notion.

The general lack of natural boundaries to mark the border of Chinese dynasties makes it difficult accurately to describe what constituted China. Except where it met with the sea, the outer limits of China changed not only from dynasty to dynasty, but also within a single dynasty. Even the reach of dynasties beyond the coast, into nearby and not-so-nearby islands, changed over time. All of these ambiguities have become highly contested issues in China's post-imperial history, with the modern government claiming all lands previously controlled by the last Chinese empire, the Qing. Qing borders were expansive, particularly after the great wars of the eighteenth century, but they were not as expansive as those of the Yuan dynasty. At least thus far, the government of China has laid no claims to the notional borders of the latter.

# Geography

Any discussion of the geography of China immediately returns us to the political problem of describing its borders. Even though it might seem a simple matter to set aside post-imperial claims to particular borders, which is beyond the scope of this book, these modern assertions of what constitutes Chinese territory rely heavily on imperial history. Dynasty after dynasty made open-ended claims on territory that were not reflected in the actual reach of their authority, and the historical maps drawn by modern cartographers balance those claims and imperial descriptions against historical accounts of what areas were actually governed. China's current government pays a great deal of attention to where the borders of imperial dynasties are drawn.

The borders of all Chinese empires included non-Chinese peoples, but highlighting this historical fact is not a political endorsement of that reality in the past or present. My purpose in describing geographic borders here is purely heuristic, though I am, of course, sensitive to the fact that those descriptions can be used to legitimize modern territorial claims. Many non-Chinese peoples were part of Chinese history, and some of them ruled China. Moreover, the borders of modern China do not and should not invoke retroactive claims to territory for earlier dynasties. The Tibetan plateau, for example, was not controlled by a Chinese dynasty for most of imperial Chinese history, nor would most of those dynasties have even imagined to have claimed it.

The brief geographic description that follows is simply an attempt to set out some rough outlines of what has constituted the territories of the many Chinese dynasties. It should not be taken as a critique of the success or failure of those dynasties to control the territory that was "rightfully" theirs by virtue of their ritual, cultural, and political position. Many modern and premodern Chinese historians have, in fact, judged dynasties by their success or failure in conquering the territory they feel rightly constitutes the cultural patrimony of all Chinese dynasties. While these sorts of territorial claims have been used to provide *casus belli* in imperial or would-be imperial campaigns of conquest, it is important to recognize that the assumptions of territorial unity, or a coherent and indivisible empire, are grounded in cultural imagination rather than consistent historical borders.

"China Proper" or, as I have preferred to call it here, "the Chinese ecumene" is formed around three main rivers, those of the Yellow, Long (Yangzi/Yangtze), and Pearl, and their tributaries. Like all of the main rivers in China, they run from west to east, emptying into the Pacific Ocean. North–south canals were often dug to connect these separate river basins, promoting unity and trade, and to allow armies to move more easily. Cultural flows followed trade routes, the most important of which were river- and canal-based. As with most pre-modern societies, imperial China relied upon water transportation to move bulk goods until the introduction of steam trains in the later nineteenth century.

The Central Plain, on the lower part of the Yellow River, was the cradle of Chinese civilization. It was the southern part of the North China Plain, the ecumene's largest alluvial plain, which was bounded to the north, west, and south by mountains—the Yanshan, Taihang, and Dabie and Tianmu—and to the east by the Yellow Sea. It was a fertile and heavily populated region, though population growth in the south began to outstrip the north by the Tang dynasty. With a brief exception at the beginning of the

Ming dynasty, the capitals of every major dynasty have been on the North China Plain. The Yellow River that ran through the middle of it was heavily silted, carrying yellow loess soil that gradually built up the river and repeatedly changed its course. Attempts to support the banks, to shore them up, or channel and control the river frequently failed with catastrophic results for the surrounding countryside.

Because it was a plain, without significant barriers to east–west movement, early Chinese politics was oriented toward problems of east–west control, keeping all of the elites on the Central Plain loyal to the imperial court. There were also fewer tributary waterways to facilitate transportation away from the Yellow River itself. These geographic features, coupled with the proximity of the northern steppe, rendered the North China Plain more convenient to horse transportation and the use of cavalry in war. Once past the mountains to the north, cavalry forces could move south relatively unimpeded through the Central Plain until the Yellow River itself. During the tenth and eleventh centuries an extensive network of canals spread across the North China Plain, connecting not only the Yellow River to the south, but also many areas across the plain to the Song capital at Kaifeng. The most important canal in that reticulated system was the Bian Canal, which connected the Yellow River with the Huai River to the south; another canal then connected the Huai to the Yangzi River further south.

The Long River (*Changjiang*) is the third longest river in the world, and the longest in Asia. Its lower part, from modern Nanjing to the coast, where Shanghai sits today, is known as the Yangzi River. Beginning on the Tibetan Plateau, it flows through Qinghai and then into Sichuan, where tributaries feed in, greatly increasing its volume. The river then moves from Hubei through the Three Gorges, and then on to the Yangzi Plain. The Han River, and several major lakes, including Dongting and Poyang, also contribute to the Long River, which finally debouches into the sea at Shanghai. Until the recent construction of the Three Gorges Dam, the river was only easily navigable for about a thousand miles from Shanghai to the Three Gorges. The Ming dynasty founder, for instance, became the dominant power in the south after a series of naval battles along that stretch of river. The Long River also marked the border between north and south China.

# Macroregions

G. William Skinner first proposed dividing China into nine socioeconomic "macroregions" in a series of articles beginning in 1964. His schema was based upon Walter Christaller's Central Place Theory that mapped market distributions systems into nested hierarchies of lesser nodes (villages, towns, or cities) feeding into progressively greater nodes, with larger populations and more extensive trade. Each macroregion had a core centered around the highest-level city in the region at the apex of the transport network. This economic system was based upon trade routes and therefore formed by geographic features that structured those routes, like rivers that facilitated trade or mountains that prevented it. These trade routes also structured social networks since the physical and economic connections between places similarly allowed or obstructed social ties. Trade routes organically formed these macroregions independently of political authority, underlining the challenge, for any would-be empire-builder, of somehow uniting functionally disparate socioeconomic units.

Skinner based his work on nineteenth-century data, and in earlier periods the organic structure of his macroregions was likely less developed. Nevertheless, the underlying concept that empires had to weld together very different regions is important for understanding imperial history. When administrative districts didn't coincide with the organic trade and social networks there was a fundamental tension, but the strategic and political interests of the government sometimes trumped ease of communication. It is also true that these macroregions were not usually separated by marked borders; peripheries faded off toward their edges where they met other regions.

One of the uses of Skinner's macroregions was to imagine China's economy as a group of economies, rather than a unified whole. This has allowed economic historians to argue that parts of imperial China's economy were not very far behind early modern Europe, while others were considerably less developed. Effectively, averaging economic performance across the empire, particularly during the Ming and Qing dynasties, presents an inaccurate description of China's economic development. I will return to this issue in Chapter 7, where I discuss imperial Chinese economic history.

## Localities

The administrative borders between districts and regions were established by dynastic governments for the purpose of governing, but the social and cultural borders developed out of historical, economic, and geographic factors. People saw themselves as living in a place or locality that was separated in some way from others. For most people in China, local identity was far more important than any identification with an imperial government or the larger concept of China itself. Even the small part of the population that traveled a significant distance from their homes—soldiers, merchants, and officials—retained strong ties to their birthplaces. That identification was cultural and administrative; not only did families explicitly perceive themselves to be from the place where their ancestors were buried, the government registered them in those places.

Just as macroregions functioned on a larger scale, the same concepts of central place theory played out on a smaller scale. People interacted with other people in their village and in the neighboring settlements connected by roads, rivers, and canals. Marriage partners were found and goods exchanged within these local areas. Imperial governments rarely reached directly down to the village level, leaving temples the most important institutions and wealthy landlords the real powers. Sometimes large temples were the main landlords, and they tended to wield their power with the same ruthless attitude as any other large landholder.

Outside the agricultural calendar, temple fairs and holidays would have marked the structure of local life. In some places, such as the Jin Shrines near Taiyuan, a temple that stood at the head of the spring which fed the local irrigation network, control over the water resources was incorporated into religious ceremonies. As Tracy Miller has shown, these ceremonies were both religious and a reflection of local power. The borders of that community reached as far as the irrigation system did, which was quite extensive, and created a strong sense of place.

Nor were the Jin Shrines unique in defining local communities. One of the most obvious indicators of local identification was the donor lists on the back of stelae, a free-standing, inscribed stone plaque, commemorating temple expansions or repairs. At more important sites, the closest government official might participate in the raising of funds for temple projects and be listed prominently. Otherwise, the list of donors ran from those who gave the most down to the lesser contributors. Local elites demonstrated their dominance by donating money and being prominently listed on temple stelae. These inscriptions also showed that many members of an elite family would each donate while, in some cases, people would band together—as a religious community or village, for example— to make a group contribution.

The borders of temple affiliation frequently overlapped, and local people did not confine their religious practice or attendance to a single institution. Stelae inscriptions show the same elite names at different temples. Patronage was expected of those elites and at least some of them contributed money throughout their local area. But how they determined which temples to donate to, or how far that patronage system reached, is unclear.

The geographic range of that patronage was only one aspect of local identification. Families also donated across generations. In some cases, it is possible to see the rise and fall of a family mixed in with the rise and fall of other families through donation inscriptions. Less wealthy people probably related to their local temples in a similar, if unrecorded, manner. Support for temples was pervasive at all levels of society, but only individuals who had the means to contribute significant amounts of money were included in inscriptions. Individuals chose to belong to a particular community, or set of communities, whose borders were clear to them.

While elite families occasionally made long-distance marriage ties, most people of any status found partners for their children within nearby communities. A family's most reliable long-term economic strategy was to consolidate power in a local area; economic and social power translated into local political authority. Elite families made marriage ties to establish, maintain, and advance their power independent of the imperial government. Local elites self-consciously created social groups that delineated them from people of lower status. The great aristocratic clans of the Tang dynasty, particularly those in Hebei, even believed themselves to be culturally superior to the imperial family. They were an extreme example of how a powerful group of families could identify themselves with a locality, and see that as more important than the imperial family's connection with the throne. Those same aristocratic clansmen also believed that their status made their members the most suitable for service in whichever imperial government happened to be in power. Their borders, in that sense, were both local and imperial. Many lesser elites during imperial history similarly based their power locally while hoping to achieve regional or even imperial power.

Merchants could also strongly associate themselves with a locality while ranging across the empire. In some periods, merchants moved across borders when China was divided, and in others they crossed borders between the empire and the outside world. Those engaged in longer-distance trade tended to follow established routes operated by people from their own home town. This provided a measure of trust since fellow townsmen knew the backgrounds of a prospective employee or fellow merchant. Even for those who traveled, social groups tied to a locality were the only hope of security and trust outside one's direct family. In the absence of enforceable contracts, a merchant needed to maintain credibility and trust to transact his business, and the only reliable way to do that was by being a known member of a community.

The ultimate acknowledgment of the centrality of local identity, and that it might trump identification with the government, was the prohibition on an official serving in his home town. It was assumed that any such official would sympathize with his family and friends, and place their interests over the government's. By definition, then, an official was forever from their home town even if they lived away for all of their adult lives.

## Conclusion

The borders of imperial China were never static, whether they were the territorial borders of the many dynasties, or the conceptual borders of the population. There was no clear and fixed "China" to attach one's loyalty to, or even a clear and fixed group of people who identified themselves as "Chinese." Rather, people lived within their own conceptual borders, primarily their locality, family, lineage, and culture. Dynastic governments claimed the people within their own imagined borders, and tried to draw together the separate socioeconomic regions under their political authority, but despite relentless rhetoric among the educated elite that demanded loyalty to Heaven's Son and his government, political loyalty was not the automatic response to finding oneself within a dynasty's borders.

Chinese borders were fluid, but there was a core area of territory and population that was unquestionably Chinese. The problem is defining accurately the borders of that core area and the culture and people who lived there. Culture changed and people migrated throughout China, while elites and the government tried to fix both culture and people in place. The topic of "China" or "Chinese history" implies an object of study with clear physical or conceptual boundaries but, in reality, no such clear borders existed.

In fact, China was much closer to Skinner's understanding of macroregions. There was a core of what constituted China, and a periphery that linked into that core. Of course, early Chinese ritual and political texts had described a very similar Chinese state of affairs in which the ruler directly controlled a central territory, with concentric rings of diminishing control. Empires did not see themselves this way, asserting absolute authority over all of the terrain within their borders no matter how far from their capital. But like the macroregions, the core of China formed organically in response to economic, social, cultural, and geographic factors. Governments imposed their own structures over these realities with varying degrees of success. The underlying borders that configured the lives of the people under those varying political authorities were always in competition with the government for control over the population. No imperial government had unfettered power to fix permanently the borders of China.

# 4

# War and the Military

War was the primary method of creating, maintaining, and destroying dynasties, but historians and political leaders seeking to claim divine sanction and moral legitimacy for a ruling house downplayed this obvious fact. Chinese leaders were able to make war serve their political purposes and establish dynasties with authority over the Chinese ecumene a half-dozen times. Even before the imperial age, war was fought on an immense scale. Dynasties consistently maintained complex and sophisticated military machines, and waged war across the whole of China. Necessity produced several significant inventions during the imperial period, including the stirrup, gunpowder, and guns. For much of its history, China had one of the most technologically, strategically, and operationally advanced military cultures in the world.

The political use of force fell on the law side of the morality-versus-law spectrum of ideal rule. Moral rulers were allowed, and even encouraged, to "punish" those resisting their correct rule, but theoretically a sage ruler would not face this choice. Unfortunately, whatever the idealized Ruist perspective suggested, China was first unified by a government, the Qin, that emphasized rule by law. Not surprisingly, they brought China together through war, rather than moral suasion. Because the Qin dynasty was short-lived and the account of its rise and fall was written in the long-lasting Han dynasty that followed, it has

always been easy to criticize Qin rulership as overly dependent upon law. That explanation of why the Qin fell so quickly
became an accepted part of Chinese political thought. Reliance
upon strict laws was not only morally reprehensible, it was the
road to destruction as well.

A strong and usually dominant strand of Chinese political
ideology assumed that the naked use of force, either in establishing or maintaining a dynasty, was only useful for brief periods
and that real, stable rule required non-violent governance. In the
Han dynasty, Lu Sheng asked the first emperor, "Your Majesty
obtained the empire on horseback, can you rule it from horseback?" Chingghis Khan received a similar comment almost a
millennium and a half later: "I have heard that one can conquer
the empire on horseback, but one cannot govern it on horseback." Civil officials did not contest the need for war to establish
a dynasty, but they were insistent that an empire could only be
run effectively and sustainably by civilian government.

Dynasties were always born of war. Their founders had to be
good generals, or at least employ good generals, to overthrow the
old order and establish a new one. Consequently, generals were
both a necessity and a threat to any political order. Those who
were loyal and effective were a precious commodity. Far more
than civil officials, the creation and maintenance of a dynasty
was dependent upon managing generals. Mismanagement could
either foment rebellion or preclude an effective response to
rebellions and invasions.

War also determined the borders of a dynasty or Chinese
state. There were, however, relatively few demarcated external
borders for Chinese empires (as discussed in the previous chapter). Not only did Chinese empires not want to acknowledge
that there were in fact specific territorial limits to their power,
they also lacked the ability to delineate and patrol exact borders.
Power tapered off quickly beyond the reach of an imperial army,
something that could be as true of lands within the empire as

those outside it. Dynastic authority, the ability to obtain resources and carry out the imperial will, was the product of direct military force or the reasonable threat of that force being used. Its absence did not necessarily immediately result in invasion or rebellion, of course, but the presence of a responsive force could dramatically alter the results in either case.

The army was also responsible for maintaining civil order. Soldiers policed the empire to apprehend criminals as well as rebels. Local magistrates were usually civil officials tasked with administering laws, civil and criminal, but in the main it would be soldiers who carried out their orders and enforced government authority. Front-line expeditionary troops did not usually serve as local military forces and vice versa. It was very difficult to maintain soldiers and officers effective in these widely disparate tasks since, for example, a capable expeditionary commander and his army would likely be generally ineffective at criminal policing. Extended periods of peace might improve the capability of an army's policing skills, and promote officers good at maintaining civil order, but undermine its battlefield effectiveness. As a result of these dual functions, states constantly struggled to balance the divergent demands on its personnel.

# Military Technology, Society, and Politics

By the beginning of the imperial period, chariots driven by aristocrats had given way to cavalry—though still without stirrups—and mass infantry armies drawn mostly from the farming population across the empire. Shock cavalry, horsemen who relied upon using the force of their mount to drive home an attack, developed in the third century, and then disappeared after the introduction of the stirrup in the fourth, a shift which remains unexplained. Mounted archery, on the other hand, was a near constant in imperial China until very late in the nineteenth century. Armies were dominated by infantry, equipped with bows, crossbows, spears, and swords. Although there were some significant shifts in weaponry over time, the distinctions in modes of warfare remained stable: mounted archers, usually steppe, and infantry, usually Chinese.

Naval warfare was equally important for commanders seeking to conquer all of the Chinese ecumene. Any attempt to create an empire like that of the Han, Tang, Song, Yuan, Ming, or Qing required crossing major rivers—the Yellow, Huai, and Yangzi, as well as many others. Crossing those rivers, and campaigning in south China as well, required a navy that could maintain riverine control long enough to transport land forces and to keep them supplied. Cao Cao, who nominally sought to preserve the Han dynasty in the early third century, was unable to move south to consolidate power after his famous defeat at the naval battle of Red Cliffs in the winter of 208–9. His initial successes had supplied him with a navy, but his commanders and soldiers were not sailors. They were defeated on the river, and then fell back to an anchorage where they chained their ships together. An attack by fire ships destroyed Cao Cao's closely packed fleet and killed many of his troops. Though he would retain power in the north, his naval defeat left China divided into three kingdoms—in the north, south, and Sichuan. Just as it had frustrated Cao Cao, the Yellow River frequently marked the farthest southern border for raiding or invading northern steppe armies because they lacked naval expertise and resources.

Another distinct and important aspect of Chinese imperial warfare was the centrality of siege engineering. Just as no army could conquer all of China without naval capability, no army could conquer much at all without the ability to capture fortified cities and, in turn, the ability to build and defend fortifications. All significant urban centers were surrounded by thick walls built with pounded earth, and some also had moats. In later times the walls were often faced with brick as well, partly to strengthen them, and partly to prevent erosion. Pounded earth walls are relatively thick for a given height, at least compared with medieval European fortification curtain walls, which were comparatively thin, high, and brittle. In Europe, fortification walls only began to resemble those in China after the introduction of the cannon. Some scholars have suggested that the nature of Chinese walls retarded the development of the cannon because the weapon did not prove nearly as effective against them. Although plausible, this was not the reason the development of the cannon slowed after the thirteenth century (a subject we shall return to later in this chapter). It is worth noting, however, that Nanjing's fourteenth-century city walls were a significant obstacle even for the Japanese army in 1937.

The most famous Chinese fortification is what, in modern times, has been called "The Great Wall." Some form of long walls and other fortifications marked the northern edge of Qin dynasty territory, but the current structure dates from the middle of the Ming dynasty (and the parts most frequently visited by tourists today are substantially late-twentieth-century reconstructions). Like many aspects of imperial China, it was more important in historiography than in history. Even for the Ming, who developed it more extensively than any other dynasty, though initially in a piecemeal fashion by local commanders, the wall was a temporary expedient that evolved into a partly effective defense against low-level threats, like small Mongol raiding parties. It was never a continuous and consistent tool of Chinese strategy for the simple reason that it couldn't solve the fundamental problem of how to defend the northern border against steppe invasion. The various long walls built at various times on the northern border, and even the Ming dynasty wall, did not mark the northern edge of the Chinese border. Functionally, northern border fortifications were at best tactical expedients, rather than strategic tools.

The shift during the Han dynasty from focusing on consolidating rule across north China to defending the northern border against steppe incursions reflected the pattern that would persist for most of imperial Chinese history. When large steppe polities consolidated under an effective leadership, like the Mongols under Chingghis Khan, they could launch large-scale invasions of sedentary Chinese states, even overthrowing and replacing them. This was relatively rare, however, and smaller steppe polities were caught between fighting other steppe polities, trading with China, working as cavalrymen for China, or raiding China for resources. When they did invade, their highly mobile armies of horse-archers made them hard to contain, but they had great difficulty capturing well-prepared, fortified positions.

## STEPPE AND SOWN

There was a persistent military, cultural, and political conflict between various nomadic and semi-nomadic groups living in the steppe, and the sedentary, agricultural Chinese for most of imperial Chinese history. Steppe people were dependent upon their horses, which they used to drive their herds from pasture to pasture as the seasons progressed. The steppe grasslands did not support a high population density, but the groups who lived there had such a facility with horses that it amplified their military potential tremendously. It was also difficult for large infantry armies to campaign in the steppe because they were neither fast enough to overtake steppe cavalry armies and their mobile families, nor were they able to carry sufficient provisions for such long expeditions. Steppe living was more precarious, however, and bad weather or being driven from a critical pasturage could impel a mobile group to seek food or resources from the sedentary Chinese population, whose agricultural system produced food surpluses that could be stored. Even in good times, most manufactured goods and luxury items were only available from China.

Chinese farming populations were also much denser and fixed in place. It was easier for imperial authorities to maintain political authority over its farmers, and to extract taxes, than for steppe leaders. Where farmers would starve if they left their crops, the mobile people of the steppe could ride away from men seeking to control them or obtain resources.

Cavalry armies were very good at attacking, but tended to withdraw when faced with superior force or the need to capture a fortified position. The only time the northern border was not a chronic military issue was when the steppe itself was politically divided, or a steppe group ruled China. Although often defeated by Chinese forces, they were always a concern.

Non-Chinese groups in the south who were outside China gradually found themselves engulfed by Chinese empires. These groups in what was, or was becoming, southern China were less organized and less mobile than northern cavalrymen. Southern indigenous populations occasionally resisted the Chinese state as it advanced into their territory, but rarely with much success. Chinese farmers migrating south gradually displaced native groups from the most productive farmland, and relentlessly ground them down through economic and demographic superiority. Unlike the northern steppe peoples, the non-Chinese in the south never presented an existential threat to the Chinese state. The greatest difficulty for Chinese armies campaigning in the far south was tropical and jungle diseases, which caused far more casualties than any battle.

## Organization

Imperial Chinese armies were extremely large in comparison to those of most pre-modern societies. Even before the imperial era, Chinese field armies had grown into the tens of thousands, necessitating the employment of skilled generals, sound logistics, and highly organized bureaucracies. Early imperial armies were usually hybrids incorporating professional soldiers alongside militia. The latter provided the bulk of armies, and the former a stiffening of cadres and the actual striking edge in battle. The Chinese ideal was the farmer-soldier, a man who labored on his farm until becoming a soldier when needed in war. Once the war was over the soldier returned to his farm, thus avoiding the need for an expensive, politically dangerous standing army. Even steppe forces, whether working for a Chinese dynasty or in the steppe itself, usually drawn from the male population of various steppe groups, were effectively militia, albeit with highly developed skills in riding and shooting. This shifted in the middle of the Tang dynasty, with the imperial army changing to a force made up of professionals, and militia confined to local defense. Their successors, the Song, also maintained a professional army, though a nostalgia for the farmer-soldiers of classical antiquity, and even the early Tang, continued to haunt government policy-makers. Repeated efforts to revive the farmer-soldier ideal were always taken seriously, despite the fact they failed to produce effective troops every time. Song statesmen feared the threat to the throne posed by a professional army, recalling the An Lushan Rebellion (755–63) that had briefly driven the Tang court from its capital and nearly destroyed the dynasty. They also lamented the cost of such an army. A farmer-soldier army, on the other hand, offered the chimera of a large, inexpensive, and loyal force able to defeat the Liao and Xixia.

Foreign rulers of China tended to keep their Chinese subjects out of military affairs, relying upon their ethnic compatriots to form primarily cavalry armies. However, a continual friction between steppe occupiers and the sedentary occupied made policing difficult for non-Chinese rulers; there was simply too much territory and too many people to suppress resistance for very long. The solution was to enlist local elites and give them a stake in the ruling dynasty's power but, as always, the difficulty was in retaining the loyalty of those empowered elites.

The Mongol rulers of China formed armies by drafting specific groups with particular capabilities: Mongols and other steppe groups supplied most of the cavalry striking forces; Muslim siege engineers working for the Mongols in the thirteenth century brought counterweight trebuchet—a swape beam with a weight on its short arm—to China; Chinese sailors working for the Mongols manned the riverine naval forces that attacked the Southern Song dynasty, and then, along with Korean sailors, tried to invade Japan; Chinese infantry fought within China, and other local forces were brought in as the Mongols rode across Eurasia. Unlike the Song army, the Mongol military did not systematically recruit men and then train them as soldiers. Unfortunately for the progress of military technology in China, they did not support the same kind of military bureaucracy that the Song had either, and the end of that bureaucracy sharply curtailed the advancement of gun technology.

After the fall of the Mongol Yuan dynasty, the Ming initially tried a new military system, permanently enrolling soldiers and officers as military families and assigning agricultural lands to them to provide for their upkeep. In theory, this would maintain a large and functional army at little or no cost to the central government's coffers; families would provide sons to the army as needed in return for their land. In practice, however, the soldiers quickly became more focused on farming, and military readiness declined sharply. So much so that, in the sixteenth century, military units composed of hereditary soldiers were completely ineffective in fighting the Wokou pirates, forcing generals to raise and train new units outside the system. Hereditary officers were no better. This would have been bad enough if the economy and the threats to the dynasty had remained static, but with changes in agriculture, society, and culture, as well as new threats forming in the steppe, the military declined in effectiveness just when a strong army was critically important. The military decline was not universal, however, and some armies and generals were episodically functional, effectively dealing with many significant threats. New Western weapons began to be adopted by the Ming in the early seventeenth century, but they were not enough to stave off growing Manchu power in the steppe.

It was the Qing dynasty that truly confronted the West and modern weaponry. In 1644 the Manchus invaded China, the

Great Wall proving an ineffective defense against large forces, and replaced the Ming dynasty. Beijing, the Ming capital, had already fallen to a bandit army before the Manchus arrived, but the imperial family and the ruling class of the Qing dynasty would all be Manchus. Some of the best Ming armies joined the Manchus against the bandits, and then continued to serve the Manchus once it was clear that the Ming dynasty was finished.

After Britain badly defeated Qing forces during the Opium War (1839–42), and the Qing military failed to stop the Taiping rebels at the beginning of the Taiping Rebellion (1850–64), halting steps

## THE OPIUM WAR (1839–42)

The Opium War between Great Britain and Qing dynasty China began because the Chinese prohibited the sale of opium by foreign, mostly British, merchants. European demand for Chinese goods—silk, porcelain, and tea—which had begun in the seventeenth century increased dramatically in the eighteenth century. The British came to dominate this trade, as tea in particular developed into a daily necessity. Lacking goods to exchange with the Chinese for these commodities, Europeans spent enormous amounts of precious metals, mostly silver, to obtain them, resulting in a significant trade deficit. In the late eighteenth century, the British began to trade opium, produced by their colonies in India, as a substitute for silver. The damaging effects of opium on the Chinese population prompted the Qing authorities to ban its trade.

Banning opium would have undermined not only Britain's now favorable trade balance with China, but also the economics of its larger international trading system. When the Qing authorities confiscated and destroyed the opium held by British merchants to enforce the ban, the British government went to war to obtain reparations and force the Qing government to permit the sale of opium. The Qing army and navy were badly outclassed, and lost a series of battles. The two sides signed the Treaty of Nanjing in 1842, which provided reparations for British merchants, and cession of the island of Hong Kong to Britain. The opium trade was officially legalized at the subsequent Treaty of Tianjin in 1858.

were taken to adopt Western arms and organization. At the very end of the imperial era, some new Qing forces were organized and armed according to Western practice, and some naval forces were likewise emulating their Western counterparts. However, the modernizing military was unable to handle mounting domestic and foreign pressures, and the Qing fell in the wake of a failed mutiny.

# Guns

One of China's most significant contributions to human civilization is the invention of gunpowder and the gun. Something like gunpowder was known as early as 808 CE and, by the late tenth century, the Song dynasty had a separate office responsible for gunpowder production in the imperial workshops. The first formulae for gunpowder were set down in writing in a military encyclopedia, *The Comprehensive Essentials from the Military Classics*, in 1044. Those three formulae required a fairly high percentage of saltpeter in their mix, though in this early stage gunpowder was used to make incendiary projectiles for trebuchet (the overall scale of production at that time is unknown). Keeping in mind that China in the eleventh century was undergoing both an institutional shift in government to a tax state, and an economic revolution toward what would be a proto-industrial revolution, it is not surprising that gunpowder production rose dramatically. By 1084, a shortfall in sulfur production required the importation of 660,000 pounds of sulfur from Japan. The scale of demand suggests large-scale production and, presumably, use of gunpowder weapons, but unfortunately the records don't allow us to determine whether this was a rare incident or part of a more regular trade relationship.

Gunpowder weapons, grenades, bombs, fire-arrows, and flame-thrower-like devices were produced in significant numbers by the beginning of the twelfth century, but were not yet effective enough to overcome the power of northern steppe groups like the Jurchen and then the Mongols. Even if they had been, the technology quickly spread across borders, though the challenges of producing sufficient saltpeter and sulfur, as well as the weapons themselves, confined most gunpowder weapons to siege and naval warfare. At least initially, gunpowder was used for its incendiary effects and the explosive potential of confining the gasses produced by burning in a container until it burst. The exact date when that same explosive force was used to propel a single projectile from a barrel remains unknown, but by the late thirteenth century the Song government was producing thousands of anti-personnel handguns. (The Chinese did not have large cannon in the thirteenth century for punching holes in fortification walls.) Although these handguns had very slow rates of fire, limited range, and were inaccurate, by the wars of the fourteenth century that established the Ming dynasty they were widely used.

Guns were invented and developed under the auspices of the Song government, during the period it transitioned first to a tax state, in the eleventh century, and then a fiscal state, in the twelfth century. (A tax state obtains more revenue from taxing non-agricultural activities than agriculture, while a fiscal state can use fiscal innovations like credit to obtain still greater revenue.) By the late fourteenth century, gun development had apparently come to a virtual halt, probably as a result of the Ming founding emperor's destruction of the fiscal state. Hongwu sought to impose a much simpler social and economic system on his empire, one which was much less dependent upon a cash economy and fiscal techniques. As with modern countries, the development and production of military technology required a government with sufficient resources and orientation to exploit the productive capabilities of its economy.

By the time the Ming dynasty shifted its economic policies in the sixteenth century, Chinese gun technology had fallen behind Europe. As Europeans established contact with China, bringing in more advanced guns, China became a recipient, rather than a producer of new gun technology. Its inferiority in military technology with respect to the West and to Japan remained a critical weakness through the end of the imperial era. Just as importantly, Manchu Qing dynasty military practice and government institutions had great difficulty incorporating European weaponry, which had been developing at an increasingly rapid rate from the sixteenth to the twentieth centuries, making adoption expensive and frustrating. Given the lag time between European developments and that technology reaching China, it was impossible to keep up. This resulted in European global military dominance in the nineteenth century, which ramified into economic superiority as well. Defeat in the Opium War, due in part to inferior military technology, has been seen in twentieth- and twenty-first-century China as the beginning of a "century of humiliation."

# Military Thought

Chinese military thought was sophisticated, pragmatic, and highly developed even before the imperial age. Most military writings were emphatically non-mystical, though there was an important strand of divination and military magic in some works, and straightforward in their advice. The most important work, "*bingfa*," translated as "The Art of War," or more recently by Victor Mair as "Military Methods," was by the mythical strategist Sunzi (also known as Sun Tzu). Although copies of *The Art of War* have been recovered from tombs dating to the third century BCE, the first "biographical" information about Sunzi is contained in Sima Qian's *The Records of the Grand Historian* (completed about 94 BCE). Sunzi's *The Art of War* and the *Master Wu* by Wu Qi (also known as Wuzi, 440–381 BCE) became the foundation texts on strategy in China. By the eleventh century CE at the latest, Chinese scholars recognized that Sunzi, or Master Sun, was a fictional military exemplar because, despite the importance of the work attributed to him, he is never mentioned in any history before Sima Qian's. Wu Qi, on the other hand, was a historical figure. A successful general in the states of Lu and Wei who later served as prime minister in Chu, he stressed the role of the general as a strategist rather than a fighter.

As in the West, it is almost impossible to connect directly military works and the prosecution of specific campaigns or battles. What is clear is that the works of Sunzi and Wu Qi were consistently studied for advice on military strategy before Sima Qian declared them exemplary strategists. Cao Cao's commentary on *The Art of War* in the third century CE began an intellectual tradition of scholars interested in military thought commenting on Sunzi. These scholars were civil officials who sought to explain various passages in Sunzi through either antiquarian clarification or historical examples demonstrating the practice of strategy. Even Cao Cao, who would go on to be one of the most famous, or infamous, warlords in Chinese history, wrote his commentary while he was still a civil official. This tradition of commentary culminated in the thirteenth century with the collection of eleven commentators into the canonical *Sunzi with Eleven Commentaries*. Unfortunately, the compiler of this work is unknown, as are his reasons for embarking on the project.

A similar process of canonization occurred in the eleventh century with the creation of *The Seven Military Classics*. Beginning in the 1030s, the Song court began to debate the necessary intellectual requirements to pass the military exam and become an army officer or military official. While the physical component, primarily skill in standing and mounted archery, was quickly resolved, the question of which texts a prospective general should study took decades to work out. Ultimately Song emperor Shenzong (1048–85, r. 1067–85) decreed what would be taught in the military academy and tested on the exam. The resulting textbook, *The Seven Military Classics*, included Sunzi and Wu Qi's works, and the eleventh-century forgery, *The Tang Taizong-Li Weigong Questions and Replies*. It established the core of Chinese military thought from then on, even though Sunzi continued to receive the over-whelming majority of attention when it came to the art of war.

## Conclusion

War and the military have been as much a part of Chinese history as anywhere else. What distinguishes imperial Chinese history even from pre-imperial history, however, is that war was repeatedly harnessed to create Chinese ecumene-spanning empires. Imperial ideology was able to incorporate war and the military into a functioning political framework flexible enough to withstand centuries of change under a single dynasty. Previous studies of imperial Chinese history have stressed the centrality of "Confucian" officials in building and maintaining these dynasties, and downplayed the use of war and the military in support of those goals. While it is certainly true that the sources for Chinese history themselves, written by civil, "Confucian," officials emphasized the role of those officials, modern historians have also generally preferred that explanation of why Chinese empires lasted for so long. From this perspective, dynasties rose and persisted because of a fundamental cultural orientation toward cohesion, not because of contingent military and political events. Chinese empires "naturally" rose, declined, and rose again.

A similar perspective was impossible for Chinese rulers, would-be rulers, and their officials. Dynasties did not arise by themselves, and political and cultural cohesion was created by hard work and a large measure of violence. A greater portion of the population participated in the military than in education, and military culture likely pervaded popular culture. This is not to say that the average farmer would not have preferred to have been able to study and become a civil official, or that military service was not harder, more dangerous, and less respectable. Rather, it is important to recognize that educated elite culture and values were necessarily not synonymous with those of the general populace. Commoners faced very different choices and understood very explicitly the state's use of the military to maintain order through violence.

Imperial Chinese armies were also consistently large, requiring the sort of bureaucracy that China became famous for. Highly developed government bureaucracies were responsible for important technological advances. For example, during the Song dynasty, it was the state's ability to collect resources and drive military innovation that led to the invention of the gun. This was an invention of the Song state, its tax system, its economy, and its army. But imperial armies could be organized in a variety of ways to support a particular ideological, ethnic, or security concern.

All successful dynasty-founding armies had to transition from conquest to maintaining order, and then defending against invasion. The half-dozen major dynasties managed to keep order and fend off outside forces for several centuries before political and military decline left them incapable of surviving. Still, military decline and political ossification were not always causally linked. They might reinforce each other, or contribute to the other, but military weakness was not the direct product of the moral decay of the imperial family and its officials.

Sometimes dynasties rose and fell simply because an army or armies succeeded or failed on the battlefield. The short-lived Chinese polities, which often only controlled parts of the Chinese ecumene, provide many examples of the failure of arms to achieve Chinese imperial ideals. There were far more failures than successes in establishing a long lasting, ecumene-spanning polity. Yet China is distinguished by the small number of successes, and the effective combination of war and politics to achieve them. Historians and unified Chinese states, including modern China, have emphasized the great successes of the Han, Tang, Song, Yuan, Ming, and Qing dynasties rather than all of the failures. This stance is, however, due to ideology rather than a balanced perspective on the nature of Chinese history.

# 5
# Discovery

China's long history of invention and exploration, like its wars, has often been downplayed. The Chinese origin of technologies, such as the stirrup, gunpowder, the compass, and many other devices, belies the notion that Chinese culture was antithetical to invention and innovation. Imperial governments welcomed advances in science and technology, and took up foreign learning and used foreign experts in fields like astronomy. In exploration, they sponsored trips abroad, often for religious purposes, welcomed news of foreign lands, and engaged in international trade.

The modern focus on technological and scientific invention reflects modern concerns. It was technology and global exploration that underpinned European imperialism and colonialism, and gave Europeans a significant military advantage over other peoples around the world. Until the middle of the nineteenth century, China was able to resist European and American attempts to force them open to trade and missionaries. Successive governments, of the Ming and then Qing dynasties, restricted where Western missionaries and merchants could live or travel to, and who they could interact with. But the Opium War (see previous chapter, pp. 65 and 68) demonstrated that the Qing military was far inferior to even a relatively small British force, which deployed some of the most advanced military technology of its time, including rockets and a steamship, against a sclerotic and poorly led Qing army and navy. In the aftermath, it seemed as if Britain's more advanced technology was the key factor in its victory, and by extension in the more general superiority of the West over China.

Western science and technology thus became a means and a marker of Western dominance. The relative technological backwardness of China in the nineteenth century and the subsequent, from the Chinese perspective, "century of humiliation" that began with defeat in the Opium War established the paradigm of a static, anti-technology China that was culturally unable to modernize as the West had. Modernity itself became a Western invention based upon the science that grew out of Ancient Greek philosophy, democracy (another Greek invention), and the Industrial Revolution. Europeans explored the world, creating international trade networks and enriching their societies. More so than any other area in modern times, technology has been the battleground for establishing the superiority of one culture over another.

This retrospective, modern interpretation, where the West came to see itself as fundamentally technologically advanced and China as fundamentally technologically backward, allowed many people to assume that China had no real science and even that its traditional culture was opposed to technology and innovation. China's reluctance to accept and adapt to superior Western technology, while at the same time insisting on the superiority of their own culture, amplified the Western sense that China's backwardness was due to its arrogance and chauvinism. Technology became synonymous with the West and modernity. The West created modern science and dominated the globe.

Chinese thinkers and scholars in the twentieth century were deeply concerned by the failure of their nation to modernize and tended, along with their Western counterparts, to look for fundamental characteristics of Chinese culture that had held them back. Japan, by contrast, rapidly modernized beginning in the late nineteenth century, and became an equal of the most developed

Western countries in the twentieth. For China, modernization was an issue of survival. The question was whether adopting Western culture and discarding its own was the only way. These struggles were reflected in how the history of Chinese science and technology was written.

The study of the history of Chinese science and technology really began in the twentieth century with scholars like Joseph Needham (1900–95), who first proposed his multi-volume book, *Science and Civilization in China*, in 1948. Needham had many collaborators, Chinese and Western, and *Science and Civilization* would grow to over two dozen volumes (Needham himself wrote fifteen), continuing to expand after his death. Although he was not the only scholar studying Chinese science and technology, his prominence and intellectual background framed what came to be called "The Needham Question." This was, "Why did modern science, the mathematization of hypotheses about Nature, with all its implications for advanced technology, take its meteoric rise only in the West at the time of Galileo [but] had not developed in Chinese civilisation or Indian civilisation?"

Like many people inside and outside China, Needham operated under Eurocentric models of development and science. Even as he was listing and bringing to light enormous amounts of Chinese scientific and technological discoveries, he struggled with the question of why they had not developed along the same paths and to the same extent as Western science and technology. Needham also operated under the modernist perspective that saw China as militarily weak, Confucianism as implacably anti-science and business, Daoists as proto-scientists, and China as inward-focused. In other words, he was a man of his time and place, and saw China as many people then did (and some still do). Yet he was relentless in bringing to light Chinese discoveries (some have even accused him of being overzealous in that regard), and demonstrating that China had science, medicine, and highly developed technology, like metallurgy, mathematics, and hydrology.

There must have been something in traditional Chinese culture that prevented those achievements from flourishing.

Modern Chinese scholars agreed with their Western counterparts. Highly educated Chinese men imagined that all of what happened in Chinese history was due, for better or worse, to the actions of highly educated Chinese men. They controlled China and what they believed their fellow citizens also believed, if at a less literate level. Scientific advancement and scientific immobility must have been due to something in the culture and ideology of the elite. Somehow educated Chinese men had managed both to invent some of the key technologies of the modern world—paper, printing, the compass, gunpowder, and guns—and then not exploit them.

The second area of Chinese failure in this modern Western perspective was China's failure to explore the world. For European, and later American, powers, exploration led to colonization, imperialism (originally a good thing, supposedly), economic development, and world domination. If Chinese ships had been capable earlier than European ships, why had they not sailed to western Europe? Why hadn't China colonized Europe or Africa? The fundamental European assumption that exploration was an objective good saw China's failure to explore and look outward as a clear sign of a cultural weakness. And indeed, Chinese people have criticized their culture for not being ocean-going well into the twentieth century. This is why Zheng He's seven ocean voyages from 1405 to 1433, ranging throughout Southeast and South Asia, have attracted so much attention, and their abrupt end so heavily criticized. But although the modern Chinese government and many Western scholars have characterized these expeditions as voyages of discovery, recent scholarship has amply demonstrated that they were military expeditions which directly intervened in local political struggles.

Of course, China did have extensive, and longstanding, connections with Eurasia, South Asia, and Southeast Asia. Those connections were, like those of the early modern West, commercial and religious. European explorers initially connected to and then, in some cases, took over pre-existing trade networks in South and Southeast Asia. Chinese overland contacts with the rest of Eurasia brought Buddhism and other religions to China, along with ideas and goods and, in turn, sent Chinese goods and ideas in the other direction. China was always, as Valerie Hansen described it, an "open empire" rather than the insular, closed-off "middle kingdom" of the early modern European imagination.

From the West's perspective, Chinese technology and science was encapsulated in the "Four Great Inventions." It went considerably beyond these four, but the concept has reflected back into modern Chinese arguments about the importance of China in world history. These four inventions also exclude anything that either did not have an impact on Europe or only counted in Western eyes as an aspect of culture, not science or technology, like tea. Imperial China's strong connections to the rest of the world that pre-existed the arrival of the West escaped Western consideration, as did the amplification of those ties as a result of European trade, like the trade between India and China.

## The Four Great Inventions

Portuguese and Spanish missionaries and traders in the early sixteenth century reported that, contrary to the previous European belief, the compass, gunpowder, and printing were longstanding technologies in China. They were not recent European inventions. Curiously, despite this early correction to European misapprehensions about the origins of these technologies, many Westerners still discount China as the source for the invention of gunpowder or, at least, the gun. Perhaps this was because it was military superiority that allowed the European powers to dominate the world. It remains inconceivable for some people that China could have invented such a critical technology and then fallen so far behind Europe.

A single book completed in 1044 links gunpowder and the magnetic compass. *The Comprehensive Essentials from the Military Classics* (*Wujing Zongyao*) is the first textual record of the use of the compass for navigation, as well as containing the first formulae (three of them) for gunpowder. During the Han dynasty, the magnetic compass was used for geomancy, only later being employed for navigation, and according to *The Comprehensive Essentials* at least, it was intended for land rather than maritime navigation. It was also, as its inclusion in a military encyclopedia indicates, seen as a piece of military technology.

From the European perspective, the compass was tremendously important as it made global navigation possible, but it had much less impact in China, where maritime trade with South and Southeast Asia did not require it. The compass first appeared in Italy in the fourteenth century, around the same time as gunpowder and guns. Despite its lesser significance in Chinese maritime navigation, the European version of the compass was adopted in the sixteenth century when it reached China via Japan. Chinese compasses were usually "wet compasses," magnetized pointers floating in water, rather than "dry compasses" used by Europeans, where the pointer was balanced in a box (sometimes known as a "box compass"). Dry compasses and magnetized needles suspended by a thread were, however, known and occasionally used in China before the sixteenth century.

While the compass and gunpowder are the most dramatic of China's inventions, in many respects paper, that most prosaic of things, has had the greatest impact on society. Paper was produced in China as early as the eighth century BCE, though there is no evidence for its use in writing until very late in the first century BCE. Its invention was subsequently attributed to Cai Lun, a eunuch and Han dynasty government official, around 105 CE. By the sixth century it was cheap and prevalent enough to be used as toilet paper, and during the Song dynasty it first saw use as money. Paper was just as important as printing for the mass production of texts, since the alternative media—silk, bamboo slats and wooden boards—were expensive or cumbersome. Despite this, the Han inscribed letters and government records on wooden boards, while pre-imperial texts used bamboo slats as far back as 1250 BCE. Paper was ubiquitous in imperial Chinese governments, whose early construction as centralized bureaucracies required what was at first a figurative, but later literal, flow of paperwork. Literate elites wrote on paper more and more over the course of imperial history, even before printing became widespread.

There were two different methods of printing in China—woodblock printing and movable type. Since even basic levels of literacy required thousands of characters, woodblock printing was usually a more efficient printing process in China and East Asia. It evolved out of the practice, carried out since at least the Han dynasty, of making rubbings from stone inscriptions: A piece of cloth or paper would be stuck to an inscription, tamped into the inscribed spaces, and then inked to produce a negative image of the inscription. At some point, first images and then texts were carved in negative on wood blocks and used to mass-produce images and texts on textiles and, later, paper. This mass production or replication of images and texts was used for religious purposes, and Buddhism was critical in promoting printing from the late sixth century on. The earliest extant woodblock prints on hemp paper are Buddhist scriptures from the seventh century.

Movable type was first produced in the eleventh century, using porcelain type. Song dynasty scientist Shen Kuo/Gua (1031–95) credited Bi Sheng, an artisan, with the invention around 1040. The individual characters were cast in clay and fired to produce type, and these were then arranged and inked before a sheet of paper was pressed on top. Bi Sheng also produced wooden movable type, but the printed results were less clear, at least initially. In any case, it was extensively used in the Ming and Qing dynasties. By the time of the Song and Jin, governments were using copper and bronze type for official documents and paper money—movable type allowed for the addition of markers for runs of bills to make counterfeiting more difficult. As it was generally more cumbersome and expensive than woodblock printing, movable type did not prove as popular, but it continued to be used for some printing. Whether either method was transmitted to Europe is unclear. Both arrived in Europe together around the fifteenth century, where movable type had clear advantages in the printing of European languages.

The magnetic compass allowed Europeans to navigate the seas to reach East Asia, but it was their advanced gun technology that allowed them to partly force open China in the nineteenth century. European imperialism more generally would simply not have been possible without superior guns. By the sixteenth century, European guns were much better than Chinese guns; by the nineteenth, they were far superior, despite China importing and adopting European firearms in the preceding centuries. The difference was so profound it seemed impossible that China had, in fact, invented gunpowder and the gun, prompting Europeans to adopt various myths attributing gunpowder to their own inventors, Roger Bacon among them. Consequently, one of Joseph Needham's original goals in shedding light on the history of Chinese science and technology was to demonstrate that gunpowder was a Chinese invention.

The military value of gunpowder seems obvious in retrospect, but the circumstances surrounding its initial discovery in the ninth century are unclear. It has frequently been asserted that all chemical or alchemical experimentation in China was part of the search for immortality by Daoists, making the accidental discovery of gunpowder in this pursuit rather ironic. While it is true that many experiments were part of this quest, others pursued chemical knowledge for both abstract and practical reasons unrelated to immortality. In the eleventh century, for example, one of the most important sources of sulfur was the process of transforming green vitriol into mordant for fixing dye in cloth. Sulfur was a byproduct of that process. Knowledge of chemistry was not confined to one goal, nor was it confined to some identifiable group of Daoists.

Harnessing the propulsive force of deflagrating gunpowder eventually led to the gun, where that force was used to launch a projectile out of the container. As with the initial discovery of gunpowder, exactly how the gun emerged during the twelfth and thirteenth centuries remains unclear. Gunpowder packed into a tube and ignited could launch a rocket or, if turned around, spray fire toward an opponent. Early fire tubes were placed on spears, near the point, making them fire spears. Pottery shards and other debris were sometimes included in the gunpowder, enhancing the effects of the spraying fire with small projectiles. When they were made of metal, the tubes proved reusable. At some point in the twelfth century, a large enough projectile or group of projectiles packed tightly in one of those metal tubes may have blocked the tube long enough for a stronger buildup of pressure. When this pressure was finally strong enough to dislodge the projectile, or projectiles, it or they would have covered a much greater distance than the flames, and with damaging force. In all likelihood, it was the discovery of this effect that led to the creation of the gun, which would be mass produced by the late thirteenth century.

Both gunpowder and the gun rapidly made their way across Eurasia. Early Chinese formulae for gunpowder had extensive and varied components, and it is unclear when exactly it was understood that the only required components were saltpeter, sulfur, and charcoal. Formulae for gunpowder that showed up in the Middle East as early as the 1290s had only the three active components, and fell very close to the maximum power mixture. The simplicity and uniformity of these formulae, along with the

Arabic word for saltpeter being "Chinese snow," suggest that thirteenth-century Chinese gunpowder had dropped the extraneous components and, through experimentation, discovered the most powerful mixture of saltpeter (75%), sulfur (15%), and charcoal (10%).

## Other Technology

The Four Great Inventions reflected early modern European perspectives on China, but there were many other areas of discovery as well. Some of those produced goods, such as silk, tea, and porcelain, which attracted Europeans to China. Chinese civil engineering and infrastructure, while not of immediate interest to Europeans, was also well-developed, with builders employing highly advanced knowledge of hydrology and construction to build bridges and locks, dig canals and irrigation systems, drain swamps, and lay out complete urban water systems. In astronomy, foreign knowledge—first Indian and later European—was empirically evaluated and adopted when it proved itself superior to previous approaches. Jesuit missionaries used their greater knowledge of astronomy to prove they offered something valuable to the imperial court, thus gaining access to the emperor and his officials. Traditional Chinese medicine evolved over time and, after a period of retreat in the face of Western medicine, has not only resumed its place in China, but also extended its reach to encompass the rest of the world. Perhaps most surprisingly, a second critical military technology, the stirrup, which was invented in China in the fourth century CE, has entirely escaped recognition. This small sample (the list is hardly comprehensive) highlights the extent to which it has been Western perspectives, in the early modern and modern periods, that have shaped perceptions of Chinese technological engagement and openness to knowledge.

## TEA, SILK, AND PORCELAIN

European explorers and merchants initially traveled to China to obtain silk and porcelain, only subsequently discovering the trading value of tea. Europe had imported silk from China since before Roman times, while porcelain first reached the continent, via the Middle East, by the middle of the fifteenth century. By the beginning of the seventeenth century, the Dutch, with their direct trade connections, were ordering over a hundred thousand ceramic pieces at a time. Silk was also being imported in large quantities despite flourishing silk production in both Italy and France.

It was at this time that tea was first brought to Europe from Japan—again, by the Dutch. Over the course of the seventeenth century, tea became fashionable among the upper classes in Europe and, eventually, England. It remained an expensive luxury good until the eighteenth century, only becoming an everyday drink in England with the production of Indian tea in the following century.

Although some pre-modern Chinese writers, like Shen Kuo, discussed technology explicitly, for most of imperial history technological developments were seen as part and parcel of ordinary life. Many of the technological discoveries, such as silk, tea, or porcelain, were imagined to have happened so far in the past that they were simply attributed to various ancient sages. Thus, Shennong, the Divine Agriculturist (a mythical, pre-historic deity), was sometimes given credit for inventing irrigation, the hoe, plow and axe, the calendar, and various aspects of Chinese medicine. The Chinese did emphasize that products like silk and tea, as well as other manufactured goods, were signs of their sophisticated culture in contrast to steppe peoples, who could only offer horses and other livestock in return for these goods. China's food supply was also more stable, due to their advanced techniques in irrigation and agriculture; when steppe herders lost livestock in periods of bad weather, they became dependent on their Chinese neighbors for survival.

Imperial China had an immense and sustained technological advantage over the surrounding communities. It did not create an overwhelming military advantage over steppe polities, even after the invention of guns, but in the economic, demographic, and cultural realms imperial China dominated. Enabled by the advantages of superior technology, imperial Chinese pursued relentless, culturally driven expansion into the non-Chinese areas while at the same time remaining open to new ideas, products, and technologies from well beyond its borders.

## Contact and Exploration

Imperial China was not only open to ideas, it also sought information and contacts with other peoples and cultures. Sometimes this was part diplomatic and military strategy. In 138 BCE, for example, the Han dynasty court sent Zhang Qian to Ferghana, in modern eastern Uzbekistan, in the hope that they would be able to make an alliance against the Xiongnu, who were attacking the empire. As it was, Zhang was captured by the Xiongnu and lived among them for ten years before returning to China. Over the subsequent centuries, Chinese dynasties would send thousands of emissaries to nearby rulers to advance diplomacy and trade, and to gather information. Those rulers in turn sent thousands of emissaries to China for the same reasons. Visiting emissaries and merchants were often brought before the emperor to be questioned about their homelands, rulers, and customs. The Chinese, of course, recorded the information about foreign lands from their own perspective, but they retained an abiding interest in the outside world.

Domestically, imperial China had direct diplomatic and ritual interests in visiting delegations. Foreigners were supposed to be drawn to China's superior civilization, which included its advanced products, and their arrival was a clear sign they recognized a legitimate Chinese emperor's virtue. In theory, the emperor would confirm a lesser ruler in their position, thus extending his legitimacy to them. The Chinese court would then provide that ruler with seals of office, which they were expected to use in correspondence with the imperial court; they would also be expected to adopt the calendar of that dynasty. The emperor could, in theory, reject a usurper or an insufficiently submissive foreign ruler, and send an army to install an acceptable candidate. In practice, however, the reach of the army and the interest of the court was limited enough to merely confirm any foreign change of ruler. While the court insisted, particularly to itself, that the emperor was Heaven's Son and the only legitimate ruler of All-Under-Heaven, Chinese statesmen also understood that they could only control foreign visitors and merchants by limiting their access to trade.

In addition to merchants and diplomatic missions, emperors sent and received many Buddhist monks from India and other lands. Some, like Xuanzong (*c*.602–64), went to India despite prohibitions on travel at that time. (Travel required imperial permission, possibly because of conflict with the Göktürks in Central Eurasia.) When he returned seventeen years later, he was welcomed back and honored even though he had violated the imperial will in leaving in the first place. Xuanzong had simply followed in the literal footsteps of Faxian (337–*c*.422), who traveled on foot to India in the fifth century in search of Buddhist texts. Faxian returned to China in 412 where he wrote an account of his travels, as well as translating and editing the works he had brought back. In addition to these Chinese monks, Indian Buddhists traveled to China bringing Buddhism and their own knowledge of specific texts. Many emperors honored these Indian Buddhist masters, establishing monasteries and translation bureaus for them.

For European Jesuits, it was their offers of superior technology and scientific knowledge that paved their way to the imperial court. Science and technology were attractive to the court because they offered an immediate, tangible benefit, as opposed to Christianity, which was just a foreign religion. The Jesuits proved that their knowledge of astronomy was superior to the existing Chinese system, which had its roots in Indian astronomy. After Father Schall von Bell (1591–1666) lost to a Chinese astronomer in a contest 1664, which led to all Jesuits being imprisoned, the discovery of mistakes in the 1670 calendar gave Father Ferdinand Verbiest (1623–88) a second chance to prove the value of Western science. Verbiest succeeded in three tests, predicting the length of a shadow at noon on a specified day, the positions of the sun and planets on a specified day, and the precise time of a coming lunar eclipse, returning the Jesuits to imperial favor. Some of the Jesuits also demonstrated their value to the emperor by forging superior cannon. A small number of Chinese officials did adopt Christianity, but Western mathematics, science, and technology were far more compelling. The limited interest in Christianity, particularly among the elites, troubled Christian missionaries and suggested to some in the West that the Chinese were only interested in Western technology and not Western culture. Later missionaries, particularly from the nineteenth century onward, focused more on converting Chinese commoners and less on elites or the imperial court.

The direction of naval expeditions was not entirely West to East, but Chinese efforts had focused on politics not proselytizing. Imperial China's most famous ocean expeditions, those of the Muslim eunuch Admiral Zheng He from 1405 to 1433, proceeded along established trade routes, projecting the power of Ming China throughout South and Southeast Asia, as well as bringing back information and goods from those areas. Zheng He's voyages were not about exploration—in some cases the routes followed had been functioning since at least the Han dynasty; instead they established or re-established political connections within or overlaying the commercial trade networks.

They did not proceed farther than the east coast of Africa because, as far as they knew, there was nothing to connect to on the other side. Certainly, Europe in the fifteenth century had nothing of commercial or religious value to offer China.

# Conclusion

China developed several technologies that had global impact, demonstrating that invention and discovery were part and parcel of Chinese culture, not antithetical to it. Moreover, China absorbed many intellectual influences from other places, as well as material goods. While Chinese merchants engaged with wide-ranging trade networks, governments sent diplomatic missions to other lands far and near for political and military purposes. Imperial governments certainly saw themselves as the center of the world, and acted accordingly, but they were also as open to useful foreign goods, intellectual and physical, as they were to the invention of useful technology.

Just as Chinese inventions like the compass, gunpowder, paper, printing, and the stirrup transformed other places and cultures, China was itself transformed by imports. Buddhism, for example, which reached China in the first century CE from India, had a profound and long-lasting effect on Chinese culture. And importing chili peppers from the New World via the European traders changed several regional cuisines to the point where today it's impossible to imagine them without chili peppers. Not all aspects of foreign cultures received as warm a welcome, however. In the late nineteenth century, European pride in their own culture frequently clashed with variable Chinese interest in things Western. The Qing government wanted to acquire Western military technology, but was reluctant to adopt Western tactics and organization. Underlying the friction was their public rejection of many things Western in the previous century, pronouncements that were made as much for domestic political consumption as foreign ears. Westerners believed their superior technology proved that their culture was also superior, and the fact that the Chinese had not produced the same innovations was seen as evidence that China was obstinately conservative and opposed to modernization.

China has never been a static culture, however, though technological shifts, knowledge of nature, and openness to change were more important in some periods than others. Their science and technology, including many globally important inventions, developed in China within the context of Chinese culture. Clearly, there was nothing inherent in imperial culture that prevented or retarded scientific exploration or rejected new technology, whatever the source. Many foreign products were accepted and incorporated into Chinese culture, and Chinese merchants and governments actively sought contact with foreign lands and peoples. Imperial Chinese culture produced critical pieces of technology in human history, and received critical pieces of technology from other places. It was just as much a part of human discovery as any other place or culture in that time.

# 6

# Religions

The religious historian Robert Campany has argued convincingly that the term "religion" should not be used to explain the phenomena commonly referred to as such with respect to Chinese practice. His wide-ranging and sophisticated argument can be reduced, for our purposes, to two points: "Religion" is a Western term, and a Western category of phenomena. The term and category, if used without, at least, consideration or challenge, are misleading when applied to China in a time where no equivalent term or category existed. The structural bias of "religion" as a term fundamentally configures not only the questions we ask and the answers we provide, but the forms of those questions and answers themselves. Nevertheless, keeping in mind the fundamentally misleading nature of this term, "religion" will be used here to describe those ritual or ceremonial practices that individuals and institutions use to inform, beseech, or satisfy the perceived requirements of other-worldly powers. This specifically sets aside the question of personal belief, a subject that the records and historical evidence do not reliably allow us to clarify. Presumably, many individuals believed in the powers that their rituals were directed toward. Others may well have practiced religious rituals cynically but, as with believers, the available evidence is not clear on who practiced sincerely and who did not. The sources are strongest on groups who wrote about their religion or left artifacts like buildings, art, or other material objects. Less literate or materially productive religions are not as well known. More generally, religions can be seen as traditions of thought, belief, and practice that from a historical perspective have been only partially preserved in texts, oral transmissions, and physical remains.

Keeping all of these caveats in mind, Chinese religion was a broad and diverse practice that involved virtually everyone in multiple, overlapping belief systems, some of which, like Ruism, also had non-religious aspects. At the very top of the political system, the emperor was, as Heaven's Son, solely responsible to Heaven for everything that took place in the empire, and even beyond it. Heaven could signal its displeasure with him through severe weather, unexpected astral phenomena, or popular unrest. The emperor engaged in extensive ceremonial practices to communicate with Heaven, placate it, and hope for favorable fortune. These ceremonial practices were planned by ritual specialists usually identified with Ruist beliefs.

Everyone, including the emperor, performed regular cycles of ceremonies for their ancestors, attempting to provide for them in the afterlife and beseech their help in return. This "ancestor worship," as it was called in the West, was a set of important ceremonial affirmations of the centrality of family identity, deeply embedded in the larger social milieu. The need for male offspring to perpetuate the ceremonies was an extension of the social values that prized boys over girls, and men over women. These values were so deeply held that it is impossible to separate the religious practices of venerating ancestors from the performance of ceremonies that reaffirmed the social structure. Whether it is appropriate to call that veneration "worship" once again turns on the definition of the terms as well as the sincerity of belief. Even an unbeliever, however, would have to perform ceremonies at the appropriate time for their ancestors if they wanted to remain a socially respectable individual.

Religions were interwoven with the Chinese political and social structure. Separating them from the fabric of Chinese culture, while sometimes necessary for the purposes of study, runs the risk of distorting their function. The religions themselves developed long histories of interaction with one another, sometimes hostile and sometimes cooperative, that make it likewise impossible to discuss their development across imperial history in isolation from one another. In 569–70, for example, Emperor Wu of the Northern Zhou ordered a debate between Buddhists and Daoists, as well as two reports on whether either religion should be adopted by the government. He emerged more favorable to the Daoists. In 574, he initiated a new debate, this time including the Ruists, and he subsequently ranked the Ruists best, followed by the Daoists, and then the Buddhists. Later that year he banned Daoism, Buddhism, and worshipping minor deities unsanctioned by the government, and ordered all Buddhist and Daoist monks to return to lay life. At least from the Buddhist perspective, this proscription only ended with Emperor Wu's death and the ascension of his son, Emperor Xuan, to the throne in 578. Most people in China did not worship or practice only one religion, and this plurality of belief highlighted the many similarities among the religions. Thus, as Stephen Teiser has shown, one of the earliest mentions of "the Three Teachings," Ruism, Daoism, and Buddhism, by Li Shiqian in the sixth century, was a reflection of both attempts to reconcile these separate religions as different approaches to the same thing, and its own strand of thinking on religion more generally.

The Three Teachings encompassed the larger part of literate Chinese religions (those practices that relied upon reading and writing sacred texts), though it is possible that the category of "popular religion" was much more broadly practiced. Popular religion was simultaneously a generic collection of religious practices that did not fit into the Three Teachings (or other scriptural traditions), and the source of a larger set of practices and beliefs that were influenced by and influenced the Three Teachings. Figures like Guan Yu, a third-century general who had evolved into the God of War by late imperial times, began their spiritual role as a figure of popular worship. Guandi, as he is usually known today, then moved from the imperial Martial Temple to become a protector deity in Buddhist temples. In this way, widely popular, powerful figures were sometimes absorbed into the frameworks of the Three Teachings.

Buddhism ultimately integrated very deeply into Chinese culture, but this was not true of every religion. Christianity, Islam, and Zoroastrianism were far less influential and integrated into the mainstream of Chinese society. The tiny population of Jews, notably in Kaifeng, a city in central Henan province south of the Yellow River, which was the capital of the Northern Song dynasty, adapted to Chinese culture so well that they were almost unnoticeable. Zoroastrianism was known into the Tang dynasty, and while it left some textual and visual traces, the effects of its presence is unclear. Similarly, Islam was an important religion among specific groups, mostly in northwestern China, farther out in Central Eurasia, and in steppe areas (though there were, at times, significant pockets of Muslims in the southeastern trading ports), but had little obvious effect on Chinese society as a whole. Early forms of Christianity were present in China in the seventh century, and it was proscribed, along with Zoroastrianism in the persecution of Buddhism in 845. The Huichang Suppression, or Great Anti-Buddhist Persecution, was Tang emperor Wuzong's effort to seize the wealth of Buddhist monasteries and prohibit the practice of foreign religion. Although some tiny Christian communities persisted through the intervening years, it was not until the sixteenth century when European missionaries arrived at the Ming court that Christianity grew again.

European missionaries converted many to Christianity without making the religion a significant force in Chinese society. Small Christian communities were established in the cities and countryside, but while missionaries at the imperial court contributed to science, technology, and mathematics, they failed to convert many important statesmen. As Western influence grew, however, so too did resentment of Chinese Christians. They were repeatedly persecuted, often violently. The Christian-inspired Taiping Rebellion (1850–64) further tainted Christianity in the eyes of the populace and the authorities. Nevertheless, Christian communities persisted through the fall of imperial China in 1912.

All of these Chinese traditions molded imperial and modern conceptions of what constituted religions. They were driven by narrow, specific concerns, mostly having to do with competition for support from the government and society. In the fifth century, for example, when Daoist communities adopted practices similar to those introduced by Buddhists, the Buddhist community responded by trying to fix Daoism as a philosophy and not a religion or "teaching." This was only partially successful, and

## THE TAIPING REBELLION (1850–64)

The Taiping rebellion, or civil war, was a Christian uprising led by Hong Xiuquan (1814–64), who believed himself to be the younger brother of Jesus Christ and, in 1851, declared himself the Heavenly King of the Heavenly Kingdom of Peace (*Taiping Tianguo*). The rebellion sought to overthrow the Manchu Qing dynasty as well as the existing Chinese social order. In its early stages, it demonstrated the profound decline of the Qing dynasty's regular army by not only defeating Qing forces in the field, but also by capturing the city of Nanjing in 1853.

Despite their early successes, the Taiping forces were unable to capture Beijing, the Qing capital, or to expand their territory to establish a stable, functional state. Internal tensions among the leaders undermined political stability, and their radical social program aroused resistance from elites and commoners at the local level. The Taipings were eventually defeated by new, locally based forces, assisted by a small contingent of foreign mercenaries.

religious Daoism flourished in its own right, but it was an early example to what would be continual struggles over the mantle of legitimate religion.

Religion was a contested category in imperial China because it held a powerful, central role in society. It provided an alternative source of authority from the government, and from local power-holders. Many lay practitioners of religion ranged broadly over this diverse landscape of institutions and ceremonies, believing in all of them, supporting all of them, and perceiving all of them as facets of a whole. That belief struggled against the exclusive believers, often ordained religious professionals, who saw their religion as the only true one. All of these perspectives on Chinese religious practice produced arguments and constructions of religion that suited their own positions. This diversity must be kept in mind in proceeding through overviews of popular religion, Ruism (Confucianism), Buddhism, and Daoism.

# Popular Religion

In many modern Chinese communities, a wide variety of spirits or deities are worshipped outside the bounds of Buddhism, Daoism, and Ruism. Depending upon how you define "popular religion," there is evidence that it has been practiced since the pre-imperial era. Master Kong famously remarked that one should "respect the ghosts and spirits, but keep them at a distance." While his exact meaning can be debated, he clearly accepted the existence of ghosts and spirits, as well as methods of interacting with them. The earliest Chinese writings—on oracle bones, turtle plastrons, or ox scapulae cracked by heating them with hot pokers in the search for divine answers to inscribed questions—were concerned with fortune telling. It remained an important area of interest throughout Chinese history, and is still widely practiced, including on websites offering various ways to predict a seeker's future. None of these approaches are connected to the Three Teachings, but all were and are prevalent.

Advocates for the Three Teachings frequently sought to suppress popular, or in their minds "heterodox," religious practice. Popular spirits who stood outside the established canons of accepted spirits were perceived as damaging to religious and social life. At a minimum, they offered alternative focuses for devotion and economic support, thus undermining community maintenance of orthodox practice. Once the popular worship of a local spirit grew enough to be noticed by the guardians of orthodox religion, that worship either had to be destroyed or incorporated into the mainstream through inclusion in official canons. Depending upon the significance of the local spirit, that incorporation might only be partially effective. A good example of this is the Sage Mother, spirit of the Jin Shrines near Taiyuan. The Sage Mother controlled the natural springs that irrigated enormous areas of farmland, making her critically important to the lives of the farmers. A number of literati attempted to fit the Sage Mother into their orthodox Ruist canon (including having her added to the government's official register of spirits and deities), satisfying other literati, but without much effect upon existing practice.

An important spirit like the Sage Mother was so central to local life that she did not require standing in an orthodox tradition to survive. Less central spirits might simply be suppressed by the authorities, usually at the behest of leaders of the orthodox traditions. Yet, local spirits and their shrines continued to arise from local concerns and beliefs, some even developing into empire-wide and ultimately orthodox spirits, adopted or co-opted by the orthodox traditions. In this sense, local interests were in tension with imperial or empire-wide interests. A local spirit might be expected to be less broadly powerful, but also easier to approach about a specific, local problem that the spirit could directly influence. Advocates for the orthodox traditions, including the imperial government, opposed any diversity of local practice that inherently defied their more universal claims to cosmology, other-worldly power, and moral order. Despite the powerful forces of orthodoxy, however, popular, local religious practice was a constant undercurrent in imperial China.

In addition to religious practices tied to specific, local concerns, there were also traditions of religious practice that spanned imperial China and came to be seen as essential to Chinese culture. Most holidays contained significant religious components, and were not directly tied to any of the Three Teachings. The New Years' Festival, Lantern Festival, Qingming (tomb sweeping) Festival, Duanwu Festival, Mid-Autumn Festival, and Double Ninth Festival, to name only a few holidays, are all popularly practiced, fundamental to Chinese culture, and have little to do with the Three Teachings. These holidays were so fundamental to Chinese culture that the orthodox traditions did not oppose them.

# Ruism (Confucianism)

Master Kong (551–479 BCE), or Confucius, was a well-known figure to whom many sayings and stories were attributed during the Warring States period. Surprisingly, the main work containing his views, *The Analects*, first appeared in the Han dynasty, as a manual of study for Han princes. Despite its late appearance, *The Analects* was an important work of Ruism in imperial China. Every educated man or woman would have been familiar with *The Analects*, and the values expressed within both reflected and described normative ideals of society. The Master Kong of *The Analects* became a great Chinese culture hero, and the Ruist tradition that he spoke to became strongly associated with his perspectives on what constituted a superior man, how and why a man should become educated, and how a state should be run, to name only a few of the many things he discussed. The Ruist ideal was a learned and morally cultivated man whose main interest in serving in government was promoting ethical rule and a prosperous society.

Master Kong was not the only great Ruist thinker of pre-imperial China, he was followed by Master Meng, or Mencius (372–289 BCE) and Master Xun, or Xunzi, (d. 238 BCE) most prominently. In imperial times, Ruist thought gradually became the foundation for imperial ideology. There were, however, many strands of Ruism, and even the mainstream Ruist thought in any given period was quite different from that of a different period. Apart from the active intellectual tradition that concerned many thinkers, there was also an important tradition of imperial ceremonial practice. Indeed, the Ru were seen as ceremonial specialists before the imperial period, and one of their main functions was planning and overseeing court ceremonies. Large parts of imperial ceremonies, like sacrificing to Heaven and Earth, were religious, though modern scholars seldom discuss this, preferring to transform Ruism into a secular system of thought called "Confucianism."

The Ruist interest in ceremony and the broader category of ritual encompassed both structured activities performed to address other-worldly powers, as well as the mundane instantiation of social and political norms through the regulation of behavior. There were correct ways to act in society and government that were based upon learning the correct ways to think. Men, for example, were superior to women, and the old deserved respect from the young; these correct views had to be reproduced and instituted in ritual performances. Correct thinking grew out of studying a core group of texts transmitted from antiquity, learning what they said and, just as importantly, what they meant. These classics were often cryptic or allegorical, requiring interpretation and commentary. The scholarly apparatus of learning and interpretation sometimes obscured the underlying or accompanying religious connection of Ruist beliefs with Heaven and the spirits.

Although there was continuous re-evaluation of Ruist texts and thought, accompanied by extensive commentary and scholarship, two thinkers during imperial history stand out: Zhu Xi (1130–1200) and Wang Yangming (1472–1529). Zhu Xi inherited and synthesized the flourishing intellectual culture of the eleventh century, particularly the work of the Cheng brothers, Cheng Yi (1033–1107) and Cheng Hao (1032–85). The Two Chengs, as they were sometimes known, connected Principles (*li*) with Heaven (*tian*), arguing that these principles linked Heaven to everything, from what we would now call nature to moral practices. The Principles established by Heaven meant that morality as defined by these Ruists was as "true" and unalterable as natural laws. Zhu Xi formalized and structured much of the diverse thinking of the eleventh century, most notably in establishing *The Four Books* (Master Kong's *Analects*, the *Master Meng*, the *Great Learning*, and the *Doctrine of the Mean*), which subsequently became the basis of Ruist and general education. This larger movement was sometimes called the Learning of Way, or the School of Principle (known in the West as Neo-Confucianism). A strict definition of those terms is difficult, but at a minimum most participants added a metaphysical basis for their ethical beliefs. Zhu connected, as the Two Chengs had, Principle with everything, including the investigation of all things, and argued that everything was made up of Substance (*qi*) organized by Principle. This is to say that the way that Principle organized Substance produced all things as well as their relationships to one another. It is unclear if this emphasis on Heaven and nature can be characterized as religious, but the Two Chengs and Zhu Xi were responding to the challenge of Buddhism and Religious Daoism.

Wang Yangming was the most influential Ruist thinker of late imperial China. He developed a different strand of thought from that of Zhu Xi, emphasizing innate knowledge over external study. Wang's perspective connected thought and action, understanding that a correct internal state of mind or heart would lead directly to correct actions. Although a deeply learned man, his emphasis on internal cultivation and action was more compelling to many educated people than abstract scholarship. Knowledge of good and evil were innate from birth and not rational. Moreover, action led to knowledge, since they were, in fact, the same. Although Wang Yangming's thought was much less formally religious than Zhu Xi's, it, like most Ruist thought, accepted the Ruist classics that made reference to religious traditions.

In the nineteenth and early twentieth centuries, both European missionaries and some Chinese reformers recast Ruist thought as "Confucianism," and defined Confucianism as a secular, rationalistic philosophy, rather than a religion. The missionaries did this in order to open the possibility of converting Ruist literati to Christianity, since Confucian principles were entirely consonant with Christian morality, and only lacked belief in Jesus. Chinese reformers tried to use Confucianism as either a Chinese religion or a core set of secular values to hold the new nation together as the Manchu Qing dynasty collapsed.

## Buddhism

Buddhism arrived in China in the first centuries of the Common Era and began to integrate into Chinese culture through translations of its main texts into Literary Chinese. It expanded over the subsequent centuries, particularly after the fall of the Han dynasty, gradually becoming a well-established part of Chinese society. Originally patronized by elites, both Chinese and steppe, it only later gained popularity among ordinary people. The many developments within Chinese Buddhism, further complicated by the episodic transmission of new *sutras* from India, produced a variety of Buddhisms in China, rather than a unified practice. Of course, the fundamental commonalities of these practices that united their traditions as "Buddhist," not least of which was the self-identification of these disparate traditions as Buddhist, allows them to be treated as a single phenomenon.

Buddhism's first and, for some traditions, greatest, flourishing period was the Tang dynasty. Tang imperial patronage amplified the general spread of Buddhism in society, and Buddhist monasteries and temples gained extensive economic power. The more general spread of Buddhism to all levels of society made it a regular part of Chinese culture, transforming both the culture and religion. Chinese Buddhism, for example, eventually created its own texts, and its own sects, like Chan/Zen. Buddhism adapted to imperial China even as it brought in new religious ideas; monks became common in Chinese society, as did the concept of monasticism. Buddhist monasteries were the first corporate institutions whose economic existence transcended individuals and families. Yet they also reinforced the existing moral, social, and political order, supporting and co-opting local, provincial, and imperial government through their wealth and religious capital. Buddhist religious institutions became important players in power struggles of all kinds. As large landowners, monasteries were wealthy, locally powerful, and often maintained their own security forces, giving them a stake and influence in political and legal matters.

Despite Buddhism's ubiquity in China after the Tang, it was always subject to attack as a foreign tradition. The Buddhists attacked the Daoists for merely copying Buddhist practices, and the Daoists attacked the Buddhists for not being Chinese. The nuances of these arguments were important for believers, particularly the professional practitioners of the respective religions, but did very little to affect the place of Buddhism in Chinese society. After all, people were generally accustomed to accepting and supporting a variety of religious practices simultaneously.

By the Ming dynasty, the imperial court began using Buddhism as a tool of foreign policy, cooperating with the Yellow Hat Buddhists in Tibet to convert the Mongols to Buddhism. This Tibetan–Mongolian connection through Buddhism took root, creating new lines of communication and control that lasted through the Qing dynasty. From the Tang dynasty onwards, Japanese and Korean Buddhists made pilgrimages to sites in China, transmitting Chinese culture along with Buddhist texts and practices back to their homes. In this way, the printing and distribution of Buddhist texts became a means of furthering Buddhism and imperial Chinese political influence. Chinese Buddhism became particularly important in and of itself after Buddhism almost completely died out in India around 1200.

# Daoism

In imperial China, there were understood to be two forms of Daoism, *Daojia* and *Daojiao*. *Daojia* is usually translated as Daoism/Taoism, and *Daojiao* as Religious Daoism. Daoism and Religious Daoism were rooted in the texts of the mythical sage/philosopher Master Lao (Laozi), *Daodejing*, the *Zhuangzi*, and a few other pre-imperial works. The pre-imperial relationship between these texts is less clear, as is their involvement in anything resembling religion. By the Han dynasty, however, they were often grouped together in an intellectual tradition or lineage. Sima Qian included a Daoist category in his biographies, which instantiated, or reinstantiated, the idea that the Daoist thinkers were equivalent to the Ruist thinkers, as well as the Legalists, militarists, and so on. Since these biographical categories emphasized traditions of thought rather than explicitly religious or spiritual beliefs, these various *jia* or lineages were seen as purely intellectual groupings. Fundamentally, Daoist thinkers understood that the functioning of the world was beyond the power of humans to control; that they could only ever effect momentary changes. The goal was to accept that reality and adapt to it in order to live as long and as comfortable a life as possible.

The Daoists could be seen as part of an intellectual rather than religious tradition in the same way that the Ruists have been portrayed purely as intellectuals. At the same time, both could also be seen as religiously, or at least spiritually, oriented, depending upon what aspects of their traditions' texts are emphasized. More clearly religious Daoist groups began to form in the second century CE with the creation of Celestial Master Daoism in 142. This group was initially confined to Sichuan, before spreading more broadly. Other Daoist groups, like the Shangqing sect begun in the third century, followed a similar trajectory. These groups were religious in every respect, and worked hard to conform to most social and political norms. Their move to the mainstream of Chinese society mostly shielded them from being attacked as a heterodox cult, though it also convinced the Buddhist clergy to condemn them as not really religious and, in the fifth century, accuse them of merely copying Buddhist practices. None of these attacks proved successful, however. At a minimum this was because Daoism, unlike Buddhism, was not a foreign tradition at odds with imperial political and social norms.

## Conclusion

Western conceptions of religion have generally obscured the reality of religious traditions in imperial China. If only Christianity qualifies as a religion, non-Christian traditions, particularly those that resisted Christian missionary efforts, were judged to be failures. Even allowing for a broader definition of religion, Western missionaries had many incentives to construct a model of Chinese culture that held open the possibility of mass conversions of the Chinese population to Christianity. This they did by emphasizing the pre-existing criticisms the Chinese religious traditions made against one another, and making Ruism an entirely secular tradition. The Western approach to Chinese religion was reified by the Communist government that took control of China in 1949. Marxist hostility to religion easily supported efforts to separate religion from the rest of Chinese society and culture as if it were an external factor.

But the traditions of Chinese religion were deeply and fully integrated into imperial Chinese culture. They were so important and influential as to be the constant subject of struggle and debate. All of the religious traditions were deeply concerned about the theological debates within and between traditions, and genuinely sought the truths they were seeking. While they may have sought different truths, or posed different questions, for their believers, Chinese religious traditions were serious undertakings.

Imperial Chinese religious traditions were real, lived traditions and practices. They marked the calendar, daily life, and the landscape. Imperial rituals worshipping Heaven and Earth were as important to the functioning of the state as the bureaucracy was, and dynastic officials knew that. No imperial state existed outside religious practice. Similarly, no family ignored the basic religious ceremonies that made sense of the world and fitted them into society. In between these two levels was a plethora of shrines, temples, and monasteries, representing the Three Teachings, as well as local spirits, and some of the smaller religious traditions such as Judaism and Islam. Imperial China was strongly religious, no less or more than anywhere else in the world.

# 7

# The Imperial Economy

The Qin establishment of imperial China came after centuries of economic change. The bronze age economy of the Shang (c.1600–c.1045 BCE) and Western Zhou (c.1045–771 BCE) dynasties had been based upon manorial estates controlled by hereditary aristocrats, alongside a smaller industrial and craft base producing goods for the elites like bronze vessels, chariot fittings, and weapons. The fall of the Western Zhou in 771 BCE began the Spring and Autumn Period (771–475 BCE), marked by the growth of cities, trade, and handicrafts. This was followed by the Warring States period (475–221 BCE), during which farmers were first allowed to own land in return for military and labor service, direct state control (no longer mediated through a local hereditary ruler), and taxes. Family farms became the fundamental building block of both the agricultural economy and political control. Political and military competition during this period required large armies, which were based upon large populations and growing food production. The larger states became more powerful by centralizing control over the economic resources of their territories, and professionalizing their governments and militaries. There was also increasing production of iron tools and weapons, improving agricultural production and expanding army size. At the same time, money came into widespread use, the power of merchants increased (because the more monetized economy and increased production of goods made them wealthier), and larger farms could hire labor to enhance productivity without the expense of more dependents. The Qin state, and the Qin imperial government, inherited all of those changes and incorporated them into its model of centralized rule. Just as importantly for the issue of economic development, they strove to promote the welfare of their subjects, for reasons both moral and realpolitik.

While the intellectual framework for the relationship of the government to the economy was mostly in place by the end of the Han dynasty, the economy itself continued to develop and change across the two millennia of imperial history. That development was neither consistent nor unidirectional. Even setting aside the idea of an objective metric by which to measure the "success" of the imperial economy, it is clear that changes in one place and time were sometimes subsequently reversed or taken in unexpected directions. Within individual dynasties, there were significant economic changes caused by many factors including technology, war, and government policy. There was no single Chinese economy or economic system across imperial history.

There were also many regional and local economies within China. These varied widely in their organization and wealth as a result of differences not only in geography, but in local cultures of commerce and trade. Thus, a unified government faced a significant challenge in ruling over different economies that were not necessarily well connected to one another. Efforts to implement universally advantageous policies were frustrated by divergent economic interests, a problem only ameliorated by the central government's limited ability to actually enact those policies across the empire. Where necessary, imperial officials tended to adapt to local conditions rather than turn the population against them.

Despite these caveats, it is possible to make a few generalizations across imperial history. Imperial governments imposed light taxes on their subjects, and did not interfere strongly in the economy. This was particularly true in the second millennium of imperial history. Consequently, they lacked the resources to invest in infrastructure or to increase the capacity of the government itself. It was only the utter failure of this approach in the nineteenth century that forced a radical shift toward a European-style approach. The Qing government could not develop an army capable of maintaining internal order and fending off the external threat of Western and Japanese imperialists. Nor could it industrialize quickly enough to catch up to the newly industrialized powers.

In the private sector, despite extensive trade networks inside and outside China that facilitated commerce, as well as money-lenders and pawnshops, financial institutions were limited in the extent and kinds of loans they made. These financial limitations, combined with cultural limitations on the structures of businesses, sharply curtailed capital formation and long-term industrial investment. Chinese businesses, including money-lenders, were either confined to families or used limited-term, cooperative agreements. In the case of the former, expansion was limited by the capacity of family members to control the business, while the latter tended to break down or dissolve over time as individual members withdrew their shares. There were no longstanding corporate businesses, beyond religious organizations, that could outgrow the vagaries of individual lives and interests.

## The State

The Chinese system of imperial government developed in an overwhelmingly agrarian economy, and that foundation left a lasting mark on imperial ideology and economic thinking. Farms, farmers, and food production were always understood to be the main source of imperial economic power. Imperial economic power was tightly connected to military power, with both based upon the vitality of the farming sector. The inaugural system under the Qin focused on food production and warfare. There was, however, an inherent tension between fighting and farming because the same men were expected to do both. Large armies required productive farms and available farmers. But men who were fighting could not be farming, or providing labor service for the state (though this was never a problem during the Warring States period, presumably because the campaigns were shorter). All adult male farmers owed the state both labor and military service as subjects of the emperor.

Chinese states also collected taxes in cash, silk, and grain from farmers. From the state's point of view, farmers ideally served as soldiers when needed, obviating the need to pay large, unproductive, standing armies, and keeping the political order stable by not maintaining greedy groups of armed men who could potentially attempt to seize power at any moment. In any case, farmers were presumed to prefer farming to fighting, and would thus only fight defensively before eagerly returning to their work. As long as the state was dependent upon the farming sector for its funding, resources, and manpower, its military actions would be limited. In the eyes of the Ru, this was a positive. But the economic limits also sharply constrained the power of the state itself, keeping government small, curtailing its reach, and leaving considerable power in the hands of local elites.

Beginning in the Han dynasty, the state found a way to escape the limitations of relying solely upon farmers for its economic and military power. Under Emperor Wudi, the state imposed taxes on salt, iron, and liquor, opening up another way to finance military activity, and expand the state's power relative to local society. After Han Wudi's death, these policies were debated at a two-day meeting, recorded by Huan Kuan in 81 BCE, called *The Discourses on Salt and Iron*. Although it is impossible to determine if the record is an accurate account of that meeting, the positions expressed by both sides appear to be consistent with the two poles of thinking on government involvement with the economy and taxation that persisted for the rest of imperial history.

*The Discourses on Salt and Iron* itself was categorized as a Ru text in a succession of imperial bibliographies. This is likely due to the slight bias in favor of the Ru position expressed in the debate. To a great extent, however, it is more asserted than obvious that the Ru came out on top. Oddly, although it provides a comprehensive discussion of some of the fundamental issues of taxation and border policy, and the text itself was transmitted intact for two millennia, it was never referred to in a court debate over taxation, government monopolies, or border policy. Without any other explanation, it may well be that because the work was not categorized as relevant to taxes or border policy, it was never considered for that purpose.

The two sides in the debate were the gentlemen and worthies—private Ru not employed as government officials, who opposed the government's taxes on salt, iron, and liquor—and the government ministers in favor of continuing those policies. The gentlemen argued that when the government relied only upon agricultural taxes, it did not launch offensive military campaigns into the steppes or attempt to expand the borders of the empire. Territorial expansion, on the other hand, caused steppe armies to invade the empire, thus tautologically justifying the taxes as necessary to defend the empire against barbarian raids. If the taxes were removed, not only would the revenue remain with the common people, but the reduction in military activity would convince the steppe barbarians to stop raiding.

In response, the ministers argued that steppe raiding happened regardless of Han dynasty military policy. Moreover, the very empire itself was built on expanding the borders. Were it not for earlier military expeditions, the empire would be smaller and more vulnerable to attacks from without. The taxes were necessary to maintain standing forces on the borders to protect the empire. Without them the common people would not be safe. As for leaving the revenues with the common people, if the government ceded control over important economic resources to commoners, some of those men might become so rich that they became a threat to the state. The economic value of industrial production was far too important politically to be left unregulated or untaxed. The ministers understood that the wealth generated by industry could be much greater than that generated by land.

Although the gentlemen and worthies are portrayed as winning the debates, Han dynasty policy did not change. Indeed, almost every dynasty in imperial Chinese history instituted taxes or monopolies on salt and liquor, as well as other products. Even dynasties that did not maintain large standing armies still found the revenue from salt taxes a critical part of the imperial budget. Under the Yuan, for example, government revenues overwhelmingly came from the salt monopoly, rising from two-thirds of the budget to four-fifths toward the end of the dynasty. State power relied upon the economic power that came from non-agricultural taxes.

Imperial ideology severely restricted state revenues, however, leading to a steady decline in government effectiveness as the population grew toward the end of the imperial period. Late imperial governments were episodically vigorous and effective, but relatively low rates of taxation and a desire to restrict the official size of the government made it impossible for the formal institutions to cope without accepted, but informal and illegal, practices. Men with official positions had to hire and pay their own staff to be able to manage their responsibilities, but this required them to collect bribes and other impositions to afford the extra, informal, staff. The system was, therefore, built upon regular and widespread corruption, rather than effective governing.

Under the Ming and Qing, tax rates were set at the beginning of each dynasty and kept unchanged for centuries. This was seen as both an act of imperial largesse bestowed upon the people and, for subsequent emperors, an act of filial piety in maintaining the policies of their predecessors. Unfortunately for the state, the tax rates were not percentages of individual income, but fixed amounts set for localities or, sometimes, families. Thus, in the former case, a county with a growing population might still pay the same amount in coins, silver, grain, or silk, over the span of centuries. In the latter case, fixed amounts lost value from inflation, and gave the state an incentive to compel families to separate into nuclear units in order to increase their total number (large, multi-generational families, though culturally preferred, provided fewer taxable units). The farm revenues of the Ming and Qing states did not, as a consequence, increase with the growing population or agricultural output.

The state's relationship to the economy was further complicated by the separation of the imperial treasury and the emperor's privy purse. Since there was no deficit financing until the nineteenth century, imperial governments had to rely upon actual revenues to function. The treasury leaned heavily on agricultural taxes, while the privy purse often collected the majority of commercial taxes. Traditionally, emperors held rights to the products of the forests, marshes, and mountains, but not farm production. Functionally, the privy purse claimed a share of all natural resources. This was quite lucrative, and often far exceeded the regular revenues of the state treasury. Emperors used this income for their own upkeep, though of course they could appropriate state revenues when so inclined, and sometimes gradually pulled more and more revenue sources into the privy purse. The emperor's personal money bolstered imperial authority and weakened the state, along with state officials. However, the privy purse also had to function as a reserve fund, and the account for military campaigns. Officials often tried to convince emperors to use it as a more general reserve fund for government purposes without expecting repayment, but wise emperors understood that the power of their purse provided them yet another lever of control over politicians.

In some dynasties, revenues were further limited by fiefs bestowed upon imperial family members, powerful individuals, and large groups that kept their own tax revenues. These fiefs also impeded the development of market networks across the empire. In the beginning of the Han dynasty, both imperial family members and important generals received fiefs that weakened state finances, undercut state control, and provided breeding grounds for rebellions. Other dynasties, particularly those with steppe ruling houses, also gave over large areas to fiefs, causing similar economic and political problems.

The imperial state also had strong cultural biases with respect to its revenue collection that worked against raising more taxes. Just as the worthies during the Han dynasty had advocated less taxation, Ru continued to argue that taxation impoverished the people. Although some educated men understood that the overall economy could expand faster than population growth by promoting agricultural and commercial development, many believed the economy was a fixed thing. Taxes, by definition, withdrew wealth from the people, leaving them poorer; therefore, the goal was to take no more than was necessary to sustain the most minimal government possible. While the state could claim all the wealth of the empire, the goal of ruling was to enrich the people and so it was better to store that wealth with the people than in the state treasuries. By late imperial times, this idea made it almost impossible to suggest increasing government revenues, and made surpluses morally suspect.

# Money

There are four economic uses of money: as a store of value, a means of exchange, a unit of account, and a standard of deferred payment.

A fifth, non-economic use is as an assertion of sovereignty. The political value of money cannot be overlooked, nor the ideological attitude of governments toward control of their money supply. Because coins were expensive to produce and relied upon limited supplies of copper, there were sound economic reasons for preventing coins from being exported. But there were also concerns about political control over the money itself as a physical manifestation of a dynasty's power. While governments took pride in the export and circulation of Chinese culture and ideas, they actively tried to prevent coins leaving the empire because that diminished their domestic supply. Other forms of currency, like rolls of silk, and bullion, did not excite the same concerns.

In the Warring States period, each state had their own money, meaning interstate commerce required that merchants engaged with a cumbersome exchange system. Coinage took a variety of forms, but the success of the Qin state saw its *banliang* coin, a round, copper, or bronze coin with a square hole in the center, prevail for the rest of imperial history. Rolls of silk were also used as currency from the Qin dynasty onward, and government salaries were often calculated in piculs (about 20 liters) of grain. Bullion did circulate, but didn't become central to transactions until the Ming dynasty. Unfortunately, a growing population and economy in late imperial times began to tighten the money supply. This was only alleviated by the importation of large amounts of New World silver, transported directly across the Pacific to the Chinese market, via the Philippines. That influx of silver, along with Japanese supplies brought in by trade, stimulated Chinese economic development. Unfortunately, it was concerns about the flow of precious metals from European states to China that stimulated the British interest in trading opium, rather than silver, in order to obtain tea, silk, and porcelain, as they sought to preserve their own money supply.

From the beginning, imperial governments understood many of the difficulties of managing the supply of money. Free coinage increased the supply, but produced lower quality coins that caused inflation and drove good coins out of circulation. Exclusive government control of minting raised the standard of coins, but also the cost of producing them. And, given the continued presence of substandard coins, the higher quality government coins were either hoarded or melted down. The ideal was an official coin good enough to stay in circulation, but not so good as to be hoarded or too expensive to produce. Despite knowing full well the problems that would be caused, imperial governments were always tempted to dilute the copper content of their coins.

Access to supplies of copper were crucial to increasing the money supply, and this varied over the course of imperial history. The greatest producer of coins was the Song dynasty, which minted 260 billion between 960 and 1127. At its peak of production around 1080, they made 6 billion coins per annum, which was the greatest output in all of imperial history. This massive production coincided with a highly reticulated market system and a proto-industrial revolution centered on the capital Kaifeng (more of which later in this chapter). Coin production declined after this peak, and then plummeted during the Yuan and Ming periods. However, Song coins remained in circulation for centuries afterward, serving as a necessary means of exchange for small, everyday transactions.

The need to produce a sufficient number of coins, and to protect the copper supply, led the Song to create zones where iron coins circulated, most notably in Sichuan and the northwest border region. In Sichuan, iron coins had been in circulation before the Song founding, and it was easier to maintain that system than replace it. On the border, iron coins served to impede the export of copper coins to the steppe, thus keeping them in China. Iron coins were larger and heavier, however, making the cash value of a pound of salt one-and-a-half times as heavy as the salt itself. The inconvenience of iron cash, combined with a separate system of salt-purchasing licenses that functioned like currency, induced merchants in Chengdu, the capital of Sichuan, to instead issue private letters of credit (*jiaozi*).

Song officials in Chengdu restricted the issuance of this new means of exchange to a group of sixteen merchants in 1005, and fixed the form of the bills. This still proved unsatisfactory to the government and merchants, and in 1024 Sichuan saw the birth of the world's first paper money. The new *jiaozi* were denominated in standard numbers of iron coins, and had fixed, three-year circulation periods. Paper money was still confined to Sichuan, reflecting its origin as a response to the particular problems of iron cash, but other paper licenses for trading in salt and tea were being used elsewhere. Even earlier, ordination certificates for Buddhist monks in the Song had fluctuated in exchange value based on market supply and demand (the limited supply and privileges of being an ordained monk meant that the market value of an ordination certificate often exceeded the price at which it was sold by the government). Certainly, by the eleventh century, imperial governments and markets had some experience of using paper documents to carry cash value. Improvements in printing techniques in the eleventh century likely also aided issuers, whether private or official, in creating reliable documents that were not easily forged.

The next important step in the development of paper money came after the Song lost control of north China to the Jurchen, a non-Chinese steppe people, ancestor to the Manchus, who then established their own Jin dynasty in 1127. As a result of extensive mining in the eleventh century, Song copper supplies were already running out before the court retreated to Hangzhou. Pressing military needs combined with the great loss of wealth from the fall of Kaifeng and the shortage of copper forced a turn to paper money. Early experiments were confined to a few small areas, but the outbreak of hostilities in 1161 forced the government to spread the new *huizi* paper money to the southeast region. It was carefully managed throughout the twelfth century, even as its use was extended into other areas, creating four monetary regions. Critically, the government mandated the use of paper money in tax payments, functionally requiring everyone to acquire paper money and making it a credible currency. This conservative management of the money supply was abandoned in the thirteenth century as military problems mounted.

Despite the vast increase in the supply of paper money, and the consequent discount of its face value, reaching seventy-five percent in the 1240s, paper replaced coins for large payments. Ingots of silver also began to play an important role in the Song economy, further diminishing the role of coins. The Mongol Yuan administration followed a similar trajectory in relying upon silver and paper money, as well as collecting taxes in silk and grain. Paper money replaced silver and bronze coins, though it often suffered from the same problems of depreciation in periods when the money supply was overinflated. With domestic demand dramatically reduced, massive amounts of bronze and silver coins were exported, to Japan and Eurasia respectively.

The first Ming emperor initially attempted to bring bronze coins back but quickly abandoned his efforts and shifted to paper money instead. He also prohibited the use of gold and silver, whose value fluctuated outside his control. Despite his insistence upon the face value of the paper money, it almost immediately traded at an eighty percent discount. By the first quarter of the fifteenth century, Ming paper money was worthless. The state slowly shifted from in-kind taxes to taxes paid in silver, but because so much silver had been exported supplies were limited and economic development slowed. That changed in the sixteenth century with increasing supplies of New World silver arriving aboard Portuguese and Spanish ships from the 1520s, and an influx from Japan beginning in the 1540s. The Ming state also returned to minting some bronze coins in 1527. As the volume of foreign silver dramatically increased after 1570, the Ming economy blossomed; China was now connected to global trade and the higher value of silver in China drew more and more traders there.

Despite economic difficulties in Europe, China did not experience a significant contraction in its silver supply through 1642. The economic downturn, often severe in parts of China during the seventeenth century, was primarily due to bad weather, bad harvests, rebellions, epidemics, and the disruptions of war. All of these factors were part of the fall of the Ming dynasty and the rise of the Qing, who continued the policies of collecting taxes in silver and relying upon silver for money. This shifted in the eighteenth century, as new sources of copper were discovered in Yunnan and Japanese copper was imported in large amounts. Bronze coins then replaced silver in the most commercialized parts of China, like the Jiangnan region.

Although Europeans substituted opium for silver in their transactions with China, causing China to have a net outflow of silver after 1827, this flow reversed thirty years later, despite record imports of opium. Silver did appreciate in value compared to bronze coins in the early nineteenth century, though this was likely due to lower-quality government coins, private minting, new private paper notes, and the improved quality of foreign silver. As European powers shifted to a gold standard in the 1870s, foreign silver flows increased. A limited amount of paper notes issued by banks made up a minor part of the money in circulation in the final decades of imperial history. Silver was the primary currency, with some bronze coins for daily transactions, when the Qing fell, and persisted well into the first half of the twentieth century.

## Markets

Formal markets were established by the government, or with its permission, and were walled off in order to control them. The government sought to collect taxes on goods by assessing the products brought to market, and to regulate weights and measures. Networks of markets developed more organically, though government initiatives in road-building, road repair, canal building, and canal and river dredging had significant, direct effects on trade networks. The connections between topographically accessible cities and towns were far stronger than between regions separated by mountains or swamps. Imperial administration generally conformed to the "natural" regions, since ease of movement created coherent districts. The other drivers of markets were resource-based, both in terms of mineral and natural resources, and different agricultural environments.

Markets in imperial China did not advance in a relentless progression along a single trajectory. Up until the Song dynasty, governments tightly controlled formal markets in cities and towns, and confined them to their designated sites. During the Song, however, the internal ward system broke down, as did confinement restrictions. This freeing up of markets, combined with the repair and expansion of waterways in north China, created a highly reticulated market system centered on the capital at Kaifeng. This northern system was connected to southern producers and markets through a canal network that allowed easy water transportation between the Yangzi, Huai, and Yellow Rivers. The increasing productivity and wealth of southern China, particularly the Jiangnan region (the area south of the lower reaches of the Yangzi River), was coupled with the concentration of wealth and political power in the capital to create what Robert Hartwell characterized as a "proto industrial revolution" in eleventh-century Kaifeng.

The population and economy of southern China began a marked increase during the Tang dynasty, when the attractions of superior agricultural productivity in the wetter and warmer climate overcame concerns about the diseases that also flourished in that environment. Wet rice agriculture allowed not only greater yields per acre, but also double cropping. While Kaifeng developed large-scale industrial operations requiring the transportation of vast amounts of iron and coal to its factories, food production in the south supported the workers involved. None of this would have been possible without the flourishing markets and transportations systems, lubricated by the massive increase in coin production. Money and markets flowed together to create an economy in the eleventh century that would not be equaled in many aspects until the eighteenth.

The Jurchen invasions of the early twelfth century, which led to the fall of Kaifeng in 1127, and the loss of north China, destroyed these markets. The northern water transport system broke down, the proto-industrial revolution ended, and north China fell behind the south economically for the rest of imperial history (and, really, up to the present). North China also suffered from repeated wars in the twelfth, thirteenth, and into the fourteenth centuries, further damaging the economic strength of the area that had been the center of Chinese civilization until that time. This only solidified the shift of cultural and economic power from north to south that began in the mid-Tang period. While China's overall economy shrank in the twelfth century, the Jiangnan region proved an exception, remaining on par with the most developed places in the world.

Yet even the relative rise in importance of Jiangnan was not absolute. In the eighteenth century, economic expansion connected markets across the empire to Jiangnan, but this trend reversed as population pressure drove immigration to more peripheral regions, developing those markets and their cities, and reducing the relative importance of Jiangnan. The empire always operated as a group of regional and local markets and economies, rather than a fully integrated whole. Where England or the Netherlands were similar in terms of their development to Jiangnan, they were countries with governments solely concerned with maintaining their own prosperity and markets. There were limits to what any imperial provincial official could or would do to promote the interests of his own area of responsibility over that of other regions, or the empire as a whole.

When Europeans sought to trade with China, most of their markets for buying and selling were initially in the south. The south produced the tea and porcelain the European traders wanted, as well as silk. It was also easier for southern Chinese markets to connect with international trade since they were already engaged with Southeast Asia. In that sense, Chinese markets were always at least partly open to maritime trade, though they were sometimes interrupted by government prohibitions. The usual reasons for such interruptions were crackdowns for the purposes of controlling trade, money, and people; attempts to stop piracy; or a desire to punish a foreign trader by excluding them.

# Conclusion

The history of the imperial Chinese economy was not one of gradual and consistent development, but rather of varied progress in different regions at different times. Government policies strongly affected the economy, as did wars, cultural orientations, and resources limitations. The great "what if" of Chinese economic history remains the question of whether the eleventh- and early-twelfth-century economic development in Kaifeng would have produced an industrial revolution, similar to that of England in the eighteenth century, if it had not been interrupted by the Jurchen invasions. Even after the Jurchen invasion, however, Jiangnan continued to achieve extremely high levels of economic development. The region's long-term progress suggests that, with the exception of the financial institutions which supported capital formation like banks and publicly traded corporations, imperial China had all of the other necessary components for modern economic growth.

There were, of course, astonishingly wealthy families who maintained their fortunes for generations; it was the successive dynastic governments that proved incapable of promoting and harnessing economic power. The Song state was certainly a tax state by the eleventh century, deriving the majority of its revenue from non-agricultural taxes, and may well have become a fiscal state by the twelfth century, but that economic power was not enough to overcome its military problems. If they'd had the military strength to maintain the dynasty in the face of outside aggression, then those policies would have justified themselves. Military failure discredited, or at least did not endorse, the Song's economic and political model.

China did not create or maintain one unified and coherent economic model during the imperial period. Agriculture changed over time and place, as did markets, transportation, and technology. The China that Europeans first encountered during the Ming dynasty was not the same China that they interacted with over the following centuries. Attempts to explain its economy foundered on the natural contradictions of trying to generalize across time and space. On the other hand, there are some products, like tea or silk, that are still dominated by Chinese production in the twenty-first century. China's modern manufacturing prowess was built upon its imperial foundation, and it is not surprising that the centers of production are where they were before.

# 8

# The Arts: Literature, Calligraphy, Painting, and Architecture

The arts were of great importance to the elites, with poetry and calligraphy in particular seen as demonstrations of character. Good writing was a sign of moral attainment, not just skill. While calligraphy was the premier visual art, painting conveyed much deeper symbolic meanings, and when composed with calligraphic inscriptions produced highly complex and refined meanings. But painting also included wall paintings inside temples that connected to religion and the common people's visual culture. Finally, architecture was the most obvious art form at all levels of society, with elaborate temples open to commoners and palaces confined to the elite. The subtleties of architecture changed significantly over time, though these aspects were known only to the most discerning.

## Literature

For most of imperial Chinese history, educated people did not write in vernacular Chinese. The literate discounted oral traditions, local cultures, and anything not connected to the long and highly developed written traditions of poetry and prose. Those traditions looked back to pre-imperial models that every educated person would have been familiar with. Han dynasty recensions of pre-imperial works, often accompanied by commentaries, formed the basis for learning how to read and write for all of imperial history. As a consequence of this, although new forms and interpretations did develop and were added to the corpus of knowledge, any educated Chinese person for two millennia shared many basic texts and attitudes about their significance.

Poetry was valued more than prose, and was usually required on civil service exams. Writing poetry was a basic skill for the educated, and while there were men (and a few women) famous for their verse, every educated person wrote poems. Consequently, the implications of fame as a poet were quite different from those in the modern West. To point out, for example, that Mei Yaochen (1002–60), one of the canonical commentators on Sunzi's *The Art of War*, was a famous poet (a pioneer in early Song poetry) would in no way diminish, and would likely enhance, his credibility as a writer on military strategy for someone in imperial China. Poetry was originally supposed to express "intent" in a formal, public manner, only taking on the function of personal emotional expression in the third century CE.

## *Poetry*

Although the familiarity of educated Chinese with multiple verse forms allowed for influence across genres, there were functionally six main forms of poetry in imperial China. The oldest was the ode (*shi*), which was based in *The Classic of Poetry* (*Shijing*), an anthology that reached a state similar to its present received version about the sixth or fifth century BCE. The poems themselves appear to have been composed between 1100 and 600 BCE. A larger collection of some 3,000 was putatively edited down to the received 305 by Master Kong himself. The myth of Master Kong's role in editing the collection shows both how central poetry was to educated Chinese culture, and the importance of ancient poetry to Master Kong. *The Classic of Poetry* retained its importance through the Qin, though in the Han dynasty it was believed that the Qin had burned it, meaning it had to be reconstructed from memory. There were three versions of *The Classic of Poetry* in the first half of the Han dynasty, but the version that would be transmitted for the rest of Chinese history until the present is *The Mao Poetry with Zheng's Commentary* (*Mao Shi Zheng Jian*). This was a fourth version of *The Classic of Poetry* transmitted via Mao Heng (third to second century BCE) with commentaries by Zheng Xuan (127–200 CE).

*The Mao Poetry* transmitted a tradition of allegorical readings of these ancient poems that read political criticism and commentary into what were likely folk songs. Poems allowed men to demonstrate their "intent," an official or court musician's perspective on political matters at court; the selection or presentation of a poem was therefore a political act. In theory, music-making and poetry arose spontaneously from people's feelings. The two art forms were also tied by the fact that they were performed at both the elite and commoner levels of society.

Popular songs had supposedly been collected by the Zhou dynasty government as a barometer of administration in pre-imperial China: A ruler could determine if his rule was benevolent by evaluating what the common people were singing or chanting. Consequently, poetry's inherently political function made it possible and even likely that any and every word in a poem was an allusion to contemporary political circumstances. Poems were also embedded in historical circumstances, requiring a solid education to understand what a given poem actually referred to and why that was important. While some modern commentators have dismissed or criticized the allegorical readings and their commentaries, educated Chinese in imperial China simply proceeded from those presumptions and composed their own works accordingly. Even if the collected poems did not, in fact, originally have intended allegorical meanings, educated

Chinese were taught that they did, and they were told what those meanings were.

While *The Classic of Poetry* recorded the odes of an official and his ruler, at least as it came to be understood, the second form of poetry, the elegiac (*sao*) form originating in *The Songs of Chu*, written between the third century BCE and second century CE and compiled by Wang Yi (d. 158 CE), records the laments of an educated man out of office. Qu Yuan (340?–278 BCE) wrote several of the seventeen works collected in *The Songs of Chu*, including "Encountering Sorrow (*Li Sao*)," a poem that attempted both to demonstrate loyalty to the Chu king and to criticize him for not distinguishing between loyal and disloyal officials. Even though Qu Yuan had been slandered by political rivals, he continued to criticize the ruler and his policies as an act of loyalty. Qu was eventually exiled to the south where he committed suicide in a final act of remonstrance by throwing himself into the Miluo River. The elegiac form "*sao*" was named after "Encountering Sorrow (*Li Sao*)," which used that particular meter.

Qu Yuan and *The Songs of Chu* were perceived to express the feelings of loyal officials trying to serve at corrupt courts. The differences between *The Classic of Poetry* and *The Songs of Chu* were usually understood in terms of their northern and southern origins, respectively. While *The Songs of Chu* likely reflects how southern musical traditions differed from those of the north, modern scholarship has stressed more widespread differences among regions and states from that period. *The Songs of Chu* also captures the state of Chu's institutional shamanism, whose values and imagery permeate the poems. Wang Yi's commentary converted all of the shamanistic imagery into allegories for conventional Ru values, which was how the poems would be read for the rest of imperial history.

Both *The Classic of Poetry* and *The Songs of Chu* contained pre-imperial forms of poetry that were incorporated into the literary, cultural, and ideological framework of imperial literati and officials, and both remained core works for educated Chinese. The first form of poetry created in the imperial era was "Music Bureau (*yuefu*) poetry," which would later be known as "ancient-style poetry (*gutishi*)." "Rhapsodies (*fu*)" or "rhyme-prose" will be discussed later in this chapter.

About 114 BCE, Han emperor Wudi (r. 141–87 BCE) established the Music Bureau (*yuefu*) to produce performances for official functions. Members of the bureau drew upon popular songs and work by famous poets, as well as creating songs themselves, and put them to music for the court. Their source material ranged over the entire empire and included foreign influences as well, representing a significant cosmopolitanism. The process of bureaucratizing that material into institutional practice soon transformed its diversity into a narrow formality suitable for watching rather than participation.

By the late first century BCE, Music Bureau poetry evolved into "Five-character old-style poems (*wuyan gushi*)." This five-character (per line) form was one of the most basic and persistent in China. Like the other early forms of poetry, whatever its apparent imagery, it was interpreted in moral and political terms. The *Anthology of Literature* (*Wenxuan*) compiled by Xiao Tong (501–31) in the early sixth century contains the *Nineteen Old Poems* (*Gushi Shijiu Shou*), which is the earliest extant collection of five-character, ancient-style poems. These latter were written at some point during the Han dynasty, but were in a mature form.

As the Han dynasty began to crumble politically in the late second century, poetry developed a new function, to express deeply held personal emotions. One of the first major proponents of this new use was Ruan Ji (210–63). Involved in government during highly uncertain and dangerous times, he shifted his ancient-style poetry away from anything that might be construed as political, focusing instead on philosophical and religious, particularly Daoist, issues. One of the renowned "Seven Sages of the Bamboo Grove" (see Chapter 1, p. 25), Ruan used the eighty-two poems of his

"Poems from My Heart (*Yonghuai Shi*)" to connect his personal feelings to the vicissitudes of life in a philosophical way.

The following century, Tao Qian (365–427) completed Ruan Ji's move away from public poetry to introspective dialogue. Tao's reputation would really be made much later, during the Song dynasty, but his statement in his autobiographical essay that he often wrote for his own pleasure, as well as to show his intentions, formally separated writing poetry from social function. While he did not entirely abandon poetry to show intention, his willingness to make it into a purely personal activity was a radical break from past practice. After all, Ruan Ji had not been able just to write what he wanted, because the creation of poetry remained in and of itself a public and political act.

As the Six Dynasties period progressed, the techniques of poetry writing became more formal (a side effect of examinations of phonology, which were caused by efforts to incorporate Sanskrit intonation from Buddhist sutras into Chinese), and increasingly erotic. Whereas Tao Qian retained intention in his poetry, as past practice required, late Six Dynasties poetry went over completely to louche emotion, partly influenced by unstable political times, and partly by the removal of allegory from love poems. Most of this poetry was written by the fairly small class of educated aristocrats, who could amuse one another with their works.

Poetry reached one of its greatest peaks in the Tang dynasty as a cosmopolitan court welcomed foreign influences, and combined northern and southern Chinese cultures. It was in the Tang that regulated verse (*lüshi*) developed and lyric (*ci*) matured. Even the conflict between intention and emotion was somewhat resolved, thanks to Kong Yingda's (574–648) formulation that outside events stimulated emotional responses which led to intentions. Another important factor in the flowering of poetry during this time was a new emphasis on government exams as a recruitment tool for officials in the late seventh century. Candidates were required to write odes and ancient-style poems, making poetry-writing part of the expanding opportunity for educated men to advance in the government and therefore bringing new men into the realm of poetry.

Regulated verse developed out of the formalized rules for poetry-writing at the Tang court in the seventh century. As a highly restricted and narrowly focused form of expression it provided a standard that later poets would rebel against. Of course, those poets were fully conversant with those standards because they had been taught them as part of their basic education. There were three forms of "new-style poetry (*jinti shi*)" in regulated verse: eight-lined, four-lined, and linked couplets.

Several of the most important Chinese poets appeared in the eighth century: Wang Wei (699–761), Li Bo [Bai] (701–62), and, most importantly, Du Fu (712–70). While new, less socially exalted men like Li Bo began producing important poetry in the eighth century, Du Fu was from a prominent family. Du Fu has generally been regarded as China's greatest poet (in some measure because he had mastered all forms of poetry) and, like so many poets, had a poor official career that caused him considerable emotional pain. Even his poetry was not highly regarded until late in the eighth century, after which he became increasingly prominent.

The final poetry form was the aria or lyric verse (*qu*), which arose in the Yuan dynasty. Arias were both independent poems sung outside a theatre performance, and also sung operatically within a drama. Like other kinds of verse, it had antecedents in earlier forms, from as far back as the Tang. Intriguingly, the aria developed during the Jin dynasty with strong influences from Jurchen culture; literati writers used the popular aria form, and it spread throughout urban society.

## Prose

It is difficult to define the category of prose in Chinese writing. Traditional scholars often mixed style, audience, and genre, while modern scholars are strongly influenced by Western ideas of writing. At a minimum, prose did not usually have a fixed rhythm, did not rhyme, and made regular use of particular grammatical particles. While it may have originally reflected spoken language, almost nothing of received early texts could be called "vernacular." Consequently, from very early times, students were taught to compose texts in a literary language that was separate from their way of speaking. The Literary Chinese (*wenyanwen*) of the imperial era was different from the Classical Chinese (*guwen*) of the great thinkers like Master Kong, but it looked back to that form, and every writer had learned to read by reading Classical Chinese works. Literary Chinese also changed over the two millennia of imperial history, with the language notably different in different times.

Chinese prose can be broken down into two stylistic categories: formal, mannered writing, and writing that stresses direct expression and subjects. Formal memorials to the throne were literary activities, and followed the prevailing stylistic norms of the time. As with any form of literary expression, the style chosen was understood to have moral and political meaning, and to connect to specific figures of authority.

In the Qing dynasty, the mannered form of writing was called "parallel prose (*pianwen* or *piantiwen*)." This was the main style for literati from the Han to the Tang dynasties. Parallel prose, as the name suggests, was primarily written in parallel couplets and authors layered in tropes, references, and figurative language. It developed out of rhapsodies or rhyme-prose (*fu*), and the technical and highly learned requirements of its form sometimes overwhelmed the authors' efforts to transmit emotion or ideas. During the Tang dynasty, some literati began to champion "ancient style (*guwen*)" prose, which they saw as more direct and powerful, unencumbered by the excessively mannered parallel prose.

Even as ancient style prose became dominant in the Song dynasty, aspects of parallel prose made a comeback. Indeed, the oscillations between these two poles of writing continued until the end of imperial China. Parallel prose offered immense artistic possibilities, both in the contrasting of lines in couplets and the ability to display erudition; ancient style prose was much less fussy and pretentious. Most Ming writers trying to make a name for themselves rejected parallel prose, but Qing writers revived it. Moreover, Ming and Qing civil-service exams required a form of parallel prose, the "eight-legged essay (*baguwen*)." A writer's choice of prose style at any given time was an act of positioning himself with respect to the style championed by the dominant group. Going along with or rebelling against current fashion was as much a political as artistic choice, and a writer might change styles at different times of his life or in different contexts. Regardless of the political and aesthetic arguments over which form was best, then, all educated men for the last five-and-a-half centuries worked hard to perfect their skills in parallel prose.

## Calligraphy

From the third century CE, calligraphy was regarded as the pre-eminent visual art in imperial China, at least in the eyes of many elite collectors of art. Chinese characters in many ways define what Chinese culture is, and the writing and presentation of those characters has always retained a particular power in Chinese society. Not only were the regional variations in characters reformed in the Qin–Han period to a unified system, but writing brushes and ink reached a mature form in the Han as well. Paper became more widely available too, making it easier and cheaper for an educated person to write or practice writing. Given Chinese bureaucracy ran on paperwork, educated men and the governments they served existed in a sea of characters.

The Qin dynasty unified characters by adopting "Small Seal Script (*xiaozhuanti*)," which was a modification of an earlier "Large

Seal Script (*dazhuanti*)" used on bronzes and other Warring States period objects. Small Seal Script then became "Clerical Script (*lishu*)" during the Han dynasty. These last two were required styles for government service, and came into general use among the educated. A further three styles of calligraphy developed during or shortly after the Han dynasty.

It was almost a century later that the most important calligrapher in Chinese history was born. Sometimes referred to as the "Sage of Calligraphy," Wang Xizhi (303–61) was collected and renowned for his work in his own lifetime. His most famous work, *The Preface to the Orchid Pavilion*, which opened a collection of poems composed at a party, was written in 353. Only copies (multiple versions for some pieces, single ones for others) of his works are now extant, but he became a model for all subsequent calligraphers, including his seventh son, Wang Xianzhi (344–86), who was considered almost as great as his father. It is worth noting, particularly since women were entirely excluded from the history of calligraphy before the twentieth century, that he was taught by Lady Wei (Wei Shuo, 272–349), a renowned calligrapher during the Eastern Jin (318–420).

The second emperor of the Tang dynasty, Li Shimin (Tang Taizong, r.626–49), was particularly fond of Wang Xizhi's calligraphy. He was also famous for his own "Flying White" style calligraphy. Like that of many other early calligraphers, Taizong's calligraphy is mostly known from rubbings of stelae inscriptions, copies of those rubbings, and even occasionally the stelae themselves. Among the other great calligraphers of the Tang, Yan Zhenqing (709–85) was considered almost on a par with Wang Xizhi. Yan had a successful political career, though he was demoted and sent away from the imperial court on several occasions for forthright criticism of policy. On one occasion, he offended the corrupt and inept Grand Councilor Yang Guozhong (d. 756) and was sent out of the capital to govern Pingyuan just as Yang's mishandling of border generals instigated the An Lushan Rebellion

(755–63). Yan saw that trouble was imminent and prepared his city for war. Unlike all the surrounding cities, Pingyuan held out, and Yan subsequently campaigned successfully against the rebels. Yang, on the other hand, would soon be killed by angry soldiers as he fled the capital with the emperor. In calligraphy, Yan was particularly known for his Model and Grass scripts, with his expression of Model script being widely copied. His statement on technique, that if one was upright, one would hold the brush upright, transformed an issue of technique into a moral position that was hard to dislodge. The felicitous association of brush position with moral rectitude, using the same word, *zheng*, made it an easy tool to admonish a student to be both moral and use the "proper" technique.

Yan Zhenqing was extremely influential in the Song dynasty, when all four of the "great masters" of calligraphy, Su Shi, Cai Xiang, Huang Tingjian, and Mi Fu, copied him. Cai Xiang, the only one of the four who was not also renowned for his painting, famously wrote to the great statesman, writer, and scholar Ouyang Xiu about the relative importance of his calligraphy to his role as a government official. Ouyang had requested that Cai write the calligraphy for the cover of Ouyang's work on epigraphy; Cai happily agreed to Ouyang's desires, while noting that he had rejected similar requests for his calligraphy at court. Powerful people had realized how good Cai's calligraphy was and would ask for him to do court-awarded inscriptions, but he argued to the emperor that since he was an official, and not a professional calligrapher, he should not be employed in such a menial role. The emperor agreed, though, of course, Cai was honored to attach his calligraphy to the work of a great scholar like Ouyang Xiu. For men like Cai Xiang, his accomplishments in calligraphy should not reduce him to being known as a mere artist.

Dong Qichang (1555–1636) was not only an important calligrapher and painter, but also, as a connoisseur, an extremely influential theorist of painting and calligraphy. For Dong, calligraphy was an ideal art because it allowed a scholar to express his true inner self through the abstract form of writing. Writing was abstract in the sense not of the specific content of the text, but the direct demonstration of the self in the crafting of the characters. Where painting allowed a representation of something real, even if used symbolically, and could be done by a professional painter, calligraphy was peculiar to the scholar's moral development. Dong's views carried great weight, and his categorization of painting (see p. 135) along with his views on calligraphy remain influential today. Many accomplished calligraphers from the Ming and earlier were neglected during the Qing dynasty because Dong did not discuss them. In many ways, Dong Qichang narrowed and shaped the construction of what great calligraphy was until well into the twentieth century. His taxonomy prevailed, creating groupings and schools, excluding women, and endorsing the value of spontaneous, southern authenticity over mannered northern professionalism.

## Painting

Although feudal rulers, and later wealthy landlords and merchants, had always employed professional craftsmen to make fine quality objects, decorations, and funerary art, it was not until after the Han dynasty that wealthy men began to produce "art" in their free time. This included poetry and calligraphy, as well as painting. The key distinction was between professionals, who sold their work for profit or as part of their employment, and amateurs, who were only concerned with manifesting their intentions or their inner moral cultivation. These putatively amateur artists were supposed to be less technically skilled than their professional counterparts, making a rougher execution of brushstroke, for example, a sign of the more refined practice of painting.

There was a close connection between the conceptions of calligraphy and painting, stemming in part from the identical tools and materials used: brushes, ink, and paper. The importance of refined connoisseurship in calligraphy was reflected in the stress on painters representing the inner truth of an object over its precise surface image. Particularly because of the later elevation of "amateur" scholar painters over professionals, at least in the writings of scholars collecting art, the techniques of verisimilitude and perspective, while known and practiced, were not highly esteemed. As with calligraphy, poetry, and prose, Chinese painting also involved demonstrating the knowledge of past masters and models. A great painter was a master within the tradition of painting.

During the Han dynasty, most painting was on walls, standing screens, and silk. Painters were employed by the government, a practice that would continue for the entire imperial period. The only Han paintings extant come from the tombs of elites, who could afford to decorate their final resting places. Han painters portrayed people, scenes from history, Daoist beliefs, nature (landscapes), and folk tales. Buddhist themes grew in importance after the fall of the Han, particularly in the north under the rule of the non-Chinese dynasties that sponsored Buddhist temples.

The most famous painter of the fourth century, and one of the first with extant works (or copies of works) is Gu Kaizhi (*c.*344–*c.*406), an amateur painter born into a scholar–official family in southern China. Gu painted both Buddhist and Daoist subjects, as well as portraits and figures. Two extant paintings are associated with him, *Nymph of the Luo River* and *The Admonitions of the Court Instructress*, the former on a Daoist subject and the latter illustrating Ruist ideas of exemplary female conduct. A good woman, for the Ru, was supposed to practice the Threefold Obedience and Four Virtues: obedience to her father, her husband, and then her son; virtue in ethics, speech, comportment, and works. While Gu's paintings on silk are extremely important, this should not obscure the great importance of wall painting in his time. Wall painting in Buddhist temples was a significant site for painters in the north and south, though with the exception of some cave temples, the rest of these works were all lost.

Around 550, Xie He (*fl.* sixth century) provided a set of criteria for evaluating and grading paintings in his "Six Laws of Painting." Like Gu Kaizhi before him, Xie was an educated man who was also a painter. By the sixth century, educated men were studying, writing about, and evaluating paintings and painters. Of course, only a very small group of elite men would have been familiar with the paintings under discussion, and the original paintings would all be lost to later generations. Because they were written about, however, and some of those essays were preserved, later generations knew what they were supposed to think about those lost works. Xie He's laws, while not absolute, provided an influential framework for the discussion of paintings in imperial China. A similarly influential discussion, attributed to Zong Bing in the early fifth century, the *Preface on Landscape Painting*, argued that a skillfully executed landscape piece could be a stand-in for real landscape, since it had the same spiritual resonance.

Elite interest in painting continued unabated with the empires of the Sui and Tang. The latter was famous for paintings of horses, perhaps a result of the dynasty's strong connection to the steppe, producing two masters of horse paintings, Han Gan (*fl.* eighth century), a court painter, and Cao Ba, a general. Landscape painting also continued to develop, notably with the addition of the new mineral colors of gold and blue-green. Yet while that sort of landscape painting was favored by professionals at court, Wang Wei, also a famous poet, was one of the first to paint monochrome ink landscapes. He was particularly renowned for his snow scenes, and retrospectively considered a founder of literati painting. His combination of Buddhism, poetry, and painting, along with his government service, placed him very firmly in the tradition of artists that began with Gu Kaizhi and continued to the end of the imperial period.

Two new traditions of painting developed after the fall of the Tang; northern landscape painting, and flower painting. Jing Hao (*fl.* 910–50), who lived in Shanxi in northern China during the Five Dynasties period, was considered the progenitor of the monumental landscape painters of the Song, originating a northern style distinct from earlier southern landscapes. Jing was also credited with writing *The Record of Brush Methods*, which set out his theory of landscape painting. Flower painting came into its own during the Five Dynasties period, developing into two strands, one more precise and professional, and one freer and associated with literati painters. The Five Dynasties was also notable for a painting, *Night Revels of Han Xizai* by Gu Hongzhong (937–75), currently known by a twelfth-century copy, in which the painter supposedly depicted contemporary people during their actual activities at a party one night.

Fan Kuan was the greatest painter of the early eleventh century, producing monumental landscapes, most famously *Travelers Among Mountains and Streams*. He was followed by Guo Xi (*c.*1020–*c.*1090), a highly educated court painter, most famous for *Early Spring*, a monumental landscape with his own distinctive style. Like Zong Bing, he argued that a properly done landscape painting could stand in for an actual journey into nature. A counter-current to these professional painters also developed, led by scholar artists like Su Shi and Mi Fu (1051–1107), among others, who argued for amateur rendering of nature to express their feelings. Fundamental to this scholar-official painting (*shidaifu hua*) was the belief that spontaneity and authentic emotion were more important than polished skill. Consequently, painters who wanted to claim to be, or appear to be, scholars would paint in a deliberately "amateur" style in order to be more authentic.

The Song was also notable for the deep engagement of several emperors with the arts. Emperors Taizong, Huizong, and Gaozong were serious calligraphers, with Huizong famous for his slender gold style. Huizong also painted, and patronized the imperial painting academy, which became a critical site for imperial legitimacy after Kaifeng was captured, along with Huizong himself, and the Song had to re-establish itself in the south.

Strong imperial patronage of the arts ended with the fall of the Southern Song to the Mongol Yuan dynasty. Wall-painting, both Daoist and Buddhist, was still important, but the Mongol court was much less interested in an imperial painting academy, or in employing the Chinese literati as government officials. Consequently, scholar-official painting was popular among the educated men who could not or would not serve the Mongol court. As a matter of identity, these amateur painters looked back to the literati of the eleventh century, who painted to express themselves rather than make a living or advance their careers. Even so, artists like Zhao Mengfu did serve the Mongols, and collected Song dynasty paintings. There were even highly skilled professional painters working in the Yuan painting academy, usually in Song style.

Ming dynasty painters inherited a wide variety of styles, and a very different political and cultural landscape. Imperial patronage emphasized different styles at different times, and many painters chose to stay away from court. The painter, calligrapher, connoisseur, and theorist Dong Qichang (1555–1636), who passed the civil service exams and served as a government official, provided the most influential understanding of early Ming painting. Apart from his great accomplishments as a painter and calligrapher, he is best known for dividing Ming painting into northern and southern "schools" of painting. The northern school was the professional school, and the southern school the amateur, with Dong changing the earlier scholar-official painting into "literati painting (*wenren hua*)." Literati painting was better than professional painting because, as others had argued before, it was an authentic expression of refined feelings.

The Ming dynasty also saw a great rise in woodblock printing, and the introduction of European painting techniques.

Woodblock printing developed rapidly in illustrated books, but the full effect of European art would not be felt until the following Qing dynasty. Qing emperors were more active patrons of art while also being quite conservative in their tastes. By this era, conservative or orthodox painting included both decorative, professional painting and literati painting, which could be produced professionally and academically. The truly exotic style was European painting, which was produced by both Europeans and Chinese artists.

# Architecture

The most recognizable aspect of Chinese above-ground architecture is its wooden post-and-beam system that supports a roof covered with ceramic tiles or thatch. For more than two millennia, many buildings, whether elite residences, temples, or palaces, were organized around a courtyard, or series of courtyards. Post-and-beam construction predates Chinese characters, and palace courtyards date from Erlitou culture, 2000–1600 BCE. Pounded or compacted earth was (and still is) used to make curtain walls for homes, compounds, and cities, as well as foundations for buildings. By the Qin dynasty, the palaces of the rulers of states were post-and-beam buildings, constructed on raised platforms of pounded earth, organized into courtyard complexes, and surrounded by pounded earth walls.

Because most above-ground architecture was wooden, very few buildings remain from before the Tang dynasty. The earliest extant timber building is Nanchan Temple, a Buddhist temple in Shanxi that dates to 782. There are also some earlier brick buildings, as well as cave-temple facades, the earliest extant brick pagoda, at Songyue Monastery, dating to 532. Various foundations of pounded earth have also been discovered throughout China, some from buildings with the locations of their post bases evident and, more frequently, the bases of city walls or towers. Fortunately, many clay and ceramic models of elite houses were buried in Han dynasty tombs, providing enough detail to establish the continuity of China's architectural tradition.

Some Han house models were multi-story towers, often guarded by figures holding crossbows and surrounded by walled compounds and farm animals. None show evidence of the roof curvature characteristic of later Chinese buildings. Because they did not rely upon trusses for support, Chinese roofs did not have to be straight and could be curved. This feature developed between the Han and Tang dynasties, though the reasons for the change are obscure. There was no particular advantage to a greater or lesser curvature, which changed over time and was visibly different in different parts of China.

Scholar of Chinese art and calligraphy Lothar Ledderose has broken down Chinese architecture into five levels: bracketing, bays, buildings, courtyards, and cities. The elaboration of brackets in Chinese buildings is distinctive. Beginning from their basic function of distributing the weight of the horizontal beam onto the vertical post, the brackets became increasingly complex. They allowed builders relying upon mortise-and-tenon joinery to create flexible structural systems that were aesthetically pleasing, able to support heavy roofs, and withstand bad weather and earthquakes. Bracketing also minimized the need for materials, since smaller pieces of carefully shaped wood substituted for simpler, larger members, though their use required highly skilled carpenters.

The space between two columns was referred to as a "bay," and all buildings had at least one. For reasons of proportion, the size of the bays increased with the number, as did the size of all parts of the building. For ritualistic reasons, important buildings usually faced south, and were wider than they were deep. Because the walls were not load-bearing, they did not become thicker as building size grew, but the size of the columns did increase commensurately. Large buildings had enormous columns.

Important buildings were usually placed around a courtyard. The main hall of a courtyard faced south, with subsidiary buildings on the east and west, and a gate or gatehouse to the south. Courtyards were completely enclosed on all sides, symmetrical, and rectilinear. More extensive residences would have multiple courtyards either deepening the complex to the rear, or in parallel. The farther back the complex went, the more intimate the space. Women in upper-class households became increasingly confined to the back courtyards over the course of imperial history. The imperial household itself was the most extreme version of this, with the emperor's personal areas restricted to women and eunuchs. Even monasteries, which had no women residents, distinguished between the more private areas of the inner courtyards, and the more public areas of the outer courtyards.

Similarly, Chinese cities were also ordered in this system of increasingly intimate, or higher status, inner boxes within boxes. Although the reality of a city plan did not always meet the idealized conception, considerable efforts were made to align imperial cities as closely as possible with those ideals. A good example of this is the Tang capital at Chang'an (modern Xi'an), which was designed during the Sui dynasty in 582. Roughly square-shaped, it covered more than 80 square kilometers, divided into a regular, rectangular grid, with major avenues running north–south and east–west. The main avenue, which was 150 meters wide, ran due north for 5 kilometers from the southern gate of the city to the southern gate of the government's administrative inner city, behind which was the imperial palace. The other major avenues, between 35 and 100 meters wide, connected to major gates in the city wall. The larger avenues created 110 wards, rectilinear, walled districts with their own gates, which would be closed at night. Courtyard complexes and other residences were established inside those wards, along with temples and other buildings. Several modifications were made to accommodate irregularities in Chang'an's actual shape and internal organization. Nevertheless, the ideal remained active throughout the imperial period. The closest city to realizing that ideal was Dadu (modern Beijing). Khubilai Khan decided to make the existing city his capital in 1271, which, after an extensive building program lasting until 1293, produced an imperial walled city, within a walled administrative city, within the larger capital walled city.

Since at least the Tang dynasty, Chinese imperial architectural forms spread to the rest of East Asia, where they were reproduced over the centuries to invoke the power of Chinese imperial culture. The ability to construct a city in this manner was as much a demonstration of real political and economic resources outside China as it was inside. Elaborate, expensive buildings were intimidating because of the very real power required to construct, maintain, and protect them.

## Conclusion

The arts of elites and dynastic governments shaped the history of imperial China in the active sense of molding behavior for cultural and political reasons, and in the retrospective sense of creating our view of that history based upon artifacts and texts. In China, both of these effects were explicitly understood to function in the arts. Emperors patronized the arts to demonstrate their own power and discernment, and to improve their historical reputation. Literati insisted upon the value of their own artistic production over that of mere professionals in order to empower those creative acts, and therefore themselves, with cultural and political significance. Art and the understanding of art was not peripheral to the exercise of power. The ability to compose poetry, write proficient essays, paint or write well, and construct impressive buildings and cities required power. Artistic performance by educated elites in imperial China was as much a sign of strength as an aesthetic pursuit.

# 9
# Popular Arts and Culture

There was considerable crossover between the elite arts and popular arts in areas like theatre, handicrafts, and religious objects. In imperial Chinese society as a whole, most of the population, including the elites, shared an enormous amount of common culture. This included myths, stories and sayings, along with common customs like Dragon Boat races, which most people knew about or took part in. These practices also changed over time, though later people assumed a certain timelessness about culture even for those things clearly tied to specific events. This culture both overlapped and sat somewhere between religion, literature, the arts, and formal histories, but was in many ways the abiding civilization of imperial China.

Elite, literate culture had much less influence on the majority of the population than the elites imagined or hoped for. In 1398, the first Ming emperor, Ming Taizu, promulgated a set of instructions for the common people, providing them with guidance on proper behavior. These "Six Instructions," which were supposed to be loudly proclaimed in every village six times a month, were actually based on the writings of the twelfth-century Ruist synthesizer Zhu Xi (1130–1200), but were not specifically Ruist in their content. Commoners were admonished to be filial to their parents, respect elders, live in harmony, teach their children and grandchildren, be content in their work, and avoid doing wrong.

The emperor prefaced the instructions with criticism of the literati who served in government, characterizing them as venal, corrupt, and immoral. By providing guidance directly to the people, he hoped to shift the responsibility for ordinary daily life down to the village and community elders, and away from the gentry.

Ming Taizu was likely one of the emperors most hostile to the literati or gentry level of society, but there was an inherent tension between the government, the literati/elites, and the broader population of commoners. The literati saw themselves, particularly from the Song dynasty on, as the moral and cultural leaders of society who served the emperor and set an example for the common people; emperors saw themselves as the arbiters of culture and morality, who led all the people, with the assistance of the elites, be they literati, aristocrats, or nobility. Most of the written sources for the past were created by the literati, sometimes working for the government. The perspective of the common people is mostly lost.

The distinction between elite and popular culture is not entirely arbitrary. The most obvious dividing line was between literacy and illiteracy. Elites read books, wrote documents, and connected to an enormous, complex culture through text. But the written–oral divide was not absolute. Only a literate man could write poetry, but many people could chant a poem once it was written. Literati painting was only available to a select group, but the murals in temples were broadly accessible. The extensive oral traditions of storytelling and theatre sometimes flowed from spoken to written word, rather than the other way around; folk stories became novels and plays. Even the most high-minded literatus lived in a much larger population of common people who might or might not respect his learning, or think that he was morally superior because he had studied. Indeed, based upon many of the tropes in popular fiction, it would be safe to say that officials were often seen as corrupt, and scholars obtuse and ineffective.

Some practices, such as foot binding, began among marginal populations, became part of elite culture, and then moved down to middle status people. First practiced among dancers, performers, and courtesans during the Five Dynasties period, foot binding emerged from the perception that small feet were sexually alluring for a woman. It entered elite practice during the Song and, despite early objections by some literati, it continued to spread throughout upper class Chinese society. By the Ming, foot binding was widespread among the gentry and literati. A girl from a respectable family had to have bound feet to get married, though, and since the practice had to begin from early childhood, it was not left to an individual's personal choice. Merchant families seeking to look respectable or to move into the ranks of the literati educated their sons and bound their daughters' feet. Even poorer people began to bind their daughters' feet to give them the chance to serve in a wealthier household and perhaps marry, or at least become a concubine of, a better-off husband.

Lower-class women did not bind their feet because it impeded their ability to work. The practice even defied repeated attempts to ban it by Qing emperors, starting in 1636, then 1638, and 1664. By the nineteenth century, all elite Chinese women had bound feet, and perhaps somewhere between forty and fifty percent of Chinese women overall. Neither an ancient practice, nor connected in any way with Ruist texts, foot binding was a late imperial practice that was part of late imperial culture. It only came to an end with the complete collapse of imperial culture in the twentieth century, when it was coupled with a sustained campaign to stop it.

# Decorative Arts

A material manifestation of culture that permeated every level and aspect of society was the decorative arts. Curiously, it was these very ordinary parts of Chinese culture that most interested foreign merchants—the basic items of everyday life that most Chinese people took for granted. The decorative arts also had the longest histories, indeed pre-histories, in Chinese culture, far beyond the two millennia of imperial times.

Because Chinese craftsmen had developed methods of mass production well before the imperial period, they produced vast amounts of goods, bronzes, jades, ceramics, and textiles. Most of these decorative arts were used as part of daily life and lost in the course of it, leaving only a fragmentary record as a result. Bronze vessels, for example, were widely used in rituals, even though they had been far more important in pre-imperial times. Craftsmen had reached a high level of expertise in bronze casting more than a thousand years before the Qin dynasty, and many exquisite ritual bronzes continued to be produced throughout imperial history. The same can be said for jade, which remains a highly prized material for the decorative arts, even in the twenty-first century. Jades were prized objects for wealthy connoisseurs, and the study of what made one valuable went far beyond just the quality of its carving to include color, transparency, and texture.

The two decorative arts best known outside China are ceramics and silk. Neolithic Chinese pots were already superbly crafted long before anything like history was written, and this ceramic tradition continues to the present day. Around the seventh century CE, potters began to produce porcelain, where the body is fired around 1,200 degrees Celsius, making it white and hard, and by the tenth century the pieces could be cast so thinly as to be translucent. One of the most recognized breakthroughs took place in the fourteenth century when "blue and white" ceramics were created. These white-bodied pieces with blue designs circulated widely both within China and throughout Eurasia. The most famous center for ceramic production was Jingdezhen, which remains one of the largest producers of pottery in the world today.

Silk holds a similarly important place in Chinese decorative arts. Weavers began to use silk filaments as early as the Neolithic age, and by the early imperial era a single government factory could employ thousands of workers. Rolls of silk were a form of currency, most produced by rural women as a way to pay taxes and generate a cash income. Imperial silk textiles were extravagantly dyed and woven with elaborate decorations. As a luxury good traded far outside China, silk was always more valuable than hemp or cotton cloth. Chinese silk was found in European tombs from the fifth or sixth century BCE, and a large trade was established with the Roman Empire. Even though the Byzantines had developed their own silk production by perhaps the sixth century CE—which eventually spread to Italy and France—European traders were very interested in obtaining silk when they first journeyed to China.

## Gardens

Gardens are a good example of an aspect of culture that also spanned elite practice and popular culture. Although private gardens built by elites were, in fact, private and walled off from the public at large, it was common practice to allow commoners in on holidays, or at other times, depending upon the garden. The gardener or caretaker might expect a tip, but it was possible for an interested person to gain access to an aspect of elite culture. Because these private gardens were partially public, they were yet another way for the elite to lay claim to cultural power. At the same time, something that was private as opposed to public had strong negative connotations in Chinese culture. Private things were private because they were not something one wanted to be seen by others. Moreover, Master Meng, the Warring States period Ruist thinker, had argued that a ruling lord's garden was a good thing if it was open to his subjects; it was only when the ruler closed off the natural resources of a garden from his subjects that they found it objectionable. For the Ruist elite, allowing access to their gardens may have allayed some of their ethical concerns.

Every imperial palace had a garden for the emperor's enjoyment, and even these were sometimes opened to the public. An imperial garden might also encompass a large game park, where the emperor could hunt, or at least view animals in a controlled environment. Early imperial gardens, like Qin Shihuandi's third-century BCE garden, contained ponds and lakes with islands modeled on the Isles of the Immortals, described by the following century's *The Classic of Mountains and Seas (Shanhai jing)* as the home of the Eight Immortals, where the elixir of immortality could be acquired. There were also high platforms that might entice female spirits to visit the emperor so that he could have sex with them. These early gardens set the model not only for subsequent imperial gardens, but for those of the elite as well. Chinese gardens were supposed to have a body of water with an island or islands, and if space did not permit a full-sized pond, then a miniature version could suffice. Rural estates also had gardens with the same features, though they tended to be much larger than urban gardens.

Urban gardens often sought to "borrow scenery" by constructing views that incorporated outside landscape features. This was all part of the effort to create multiple views and perspectives within the garden. The space could thereby remain private while seeming expansive. Sima Guang's *Garden of Solitary Delight* incorporated the views of nearby mountains. Pathways wound around the garden and the water within it, preventing unobstructed views of the whole. There were also numerous architectural features, commensurate with the size of the garden. As with Chinese landscape painters, garden designers placed buildings and pavilions within their works to show that the environment was welcoming, not wild; nature was to be enjoyed in a controlled and civilized manner.

The poet and statesman Bo/Bai Juyi (772–846) suggested that one sixth of the land of a residence should be given to the house, half to lakes or ponds, and a third to stands of bamboo. His breakdown shows that the function of a garden went beyond decoration. Particularly for an urban garden, a reliable supply of water was extremely important, as was ready access to bamboo, useful for construction and fuel. As Chinese art historian Craig Clunas has shown, until the late Ming dynasty gardens were supposed to be productive as well as beautiful.

In addition to water, gardens were supposed to have hills (standing in for mountains) and rocks. The first known rock garden dates to the Western Han, but in the Tang dynasty individual rocks themselves came to be objects of connoisseurship. The most famous came from Lake Tai, with the ideal rock having holes made by erosion. For their gardens, scholars sought extraordinary-looking rocks that would be admired for their own sake. The desire for peculiar-looking rocks extended to very large boulders, which wealthy men and emperors spent fortunes on transporting to their gardens. Song emperor Huizong famously had a series of bridges over the Grand Canal broken to allow large rocks to be floated up from the south for his garden. Wealthy men in the Ming paid to break the walls over city gates to move similarly enormous rocks into their urban gardens.

Like landscape painting, with which gardening had a dynamic relationship, the choice of plants was deeply infused with meaning. Scholars always liked to imagine themselves to be good men who suffered the unjust political winds in order to serve, or try to serve, in government. Thus, the "Three Friends of Winter," the pine, bamboo, and Chinese plum, stayed green throughout the winter, suffering adversity without dying. This sort of symbolism carried over into other kinds of painting as well. Even imperial gardens used symbolic plants to evoke the scholar's garden, allowing emperors to depict themselves as simple scholars.

Imperial gardens could also hold more elaborate buildings and gardens within gardens. The summer palace in Beijing had European-style buildings designed by Jesuit missionaries. It even had a full-sized model of Shantang Street in Suzhou, a southern city, and a copy of another garden from the Jiangnan region which had impressed the Qianlong emperor when he had visited. The importance of the summer palace to the emperor made it a target for European armies during the Second Opium War in 1860. Europeans believed that the emperor was insulated from the severity of their threats, so they destroyed the summer palace to make him personally aware of their power.

A private garden forms one of the main sites of the great Chinese novel *Dream of the Red Chamber*, the story of two great and wealthy branches of a family who live in adjacent compounds. The garden, which is constructed for the visit of a member of the family who has become an Imperial Consort, is described in some detail, and different vantage points are named by the characters. A critique of elite life, the novel was widely read among the literate populace, and admired for its intimate portrayal of highly refined, if fundamentally corrupt, wealthy families.

### DREAM OF THE RED CHAMBER

*Dream of the Red Chamber* or *The Story of the Stone* is a novel written in the mid-eighteenth century by Cao Xueqin. The original manuscript with eighty chapters circulated under a number of titles before it was printed in 1791, with an additional forty chapters added by Gao E. It is generally considered to be the greatest Chinese novel ever written. The novel recounts the decline of two branches of the Jia family, who live in neighboring mansions in Beijing. With several dozen main characters and hundreds of minor characters, it portrays a vast array of human interactions, and sympathetic and unsympathetic characters, amid a growing sense of refined malaise that ultimately leads to the family's destruction.

## Public Performance and Theatre

There were public performances of many kinds in imperial China, some of which fell into the category of theatre. Theatre itself moved from the popular realm into the literati realm in late imperial China, but still retained its popular attractions. Many aspects of public performance began before the imperial era, and carried over into imperial times. Some of these events were more public, that is to say, open to commoners, than others. Royal and later imperial sacrifices were not open to the public; participation being confined to members of the court. Sometimes the procession to the sacrificial site passed through the streets of a city, but barriers could be in place to prevent public viewing. Other ceremonies in local, non-imperial temples would have been open to the public. In small communities, those ceremonies would have been central to village life and included commoners.

Commoner men enrolled in the army or militia would have taken part in military training, drills, and army ceremonies. Until well into the Tang dynasty most adult men would have participated in rudimentary military training, including group drills subject to regular inspection by a local leader or government official. Training in arms was, however, a double-edged sword for governments. It was extremely useful to call up the mass of farmers for military service when needed. Unfortunately, that same population of trained men could be used against the government. The solution was to only distribute arms when there was a military necessity, and to substitute wrestling for armed military training during times of peace. Wrestling was promoted among the populace because there was a low likelihood of death or serious injury, and it ensured at least some men would learn to fight. Competitions were easy to stage and became one of the most reliable forms of popular entertainment.

The other kind of martial performance frequently discussed by Master Kong and the Ru was the village archery contest.

This was a formalized gathering where men performed archery and drank wine. In Ru theory, it provided an opportunity for the affirmation of hierarchical relationships in society; where younger men deferred to older men, social inferiors deferred to social superiors, and polite competition brought everyone together. Although the few early mentions of what actually happened in such contests paint a less seemly picture, it remained a Ru ideal. Dozens of archery manuals were written during the imperial period focusing, we are told (they are no longer extant), on the village archery contest. Imperial governments were, however, often reluctant to allow anyone, commoner or elite, to own weapons.

During the second half of the Song dynasty, however, archery societies sprang up in southern China. These groups usually practiced at temples, since they had the only open spaces large enough for archery practice. In northern China, because there were more frequent conflicts with steppe groups, archery and other martial arts were more common. The Song government allowed and even promoted these practices because it hoped a more martially capable citizenry would strengthen it.

Non-Chinese groups had their own public displays of martial arts that were somewhat different from the Chinese. Steppe groups were even more interested than their neighbors in wrestling and archery, though their more formal archery contests were not infused with Ruist ideology. For steppe people, archery was directly tied to their identity, as was riding horses. Men, and sometimes women, took part in large-scale hunting expeditions, which helped cement social and political ties, as well as training men in cooperative activities that also had military applications. Wrestling was also extremely popular, and in China, steppe or Türkic people were considered the best wrestlers. All of these activities allowed the public display of skills that demonstrated membership in the community, and might allow for an improvement in personal status.

Wrestling had a long history before the imperial era, and the Qin dynasty continued to promote wrestling competitions to demilitarize the population. Though already quite popular in both elite and commoner circles, these competitions fed into the Han dynasty "Hundred Events," which was a more general set of public performances that included dancing, acrobatics, singing, and some of the other precursors to theatre. The Hundred Events took place on certain holidays or to celebrate something important; sometimes they were put on when important foreign dignitaries were visiting. At the same time, all of the various activities were also performed throughout the year in different venues from the very public to the inner palace.

If we consider the contents of modern Chinese cinema a public performance, all aspects of the Hundred Events continue up to the present. The more formal development of theatre seems to be tied to performances in rural temples, and grew out of sacrifices and ceremonies involving dance. At a minimum, there is solid archeological support for the existence of stages at temples, beginning in the eleventh century. Of course, many earlier stages may have either been destroyed or dismantled because they were temporary structures established as needed for performances. Most extant stages are oriented toward the temple of the patron deity of the site so that the gods could enjoy the performance. The setting did not restrict performances to religious topics or didactic morality plays, however; Chinese deities apparently had similar tastes in performance to ordinary mortals.

The importance of rural performances in the development of theatre, highlights the wide regional and local variation in public performances. Chinese empires encompassed such vast territories which were only somewhat connected to one another by roads and waterways that local cultures abounded. From the perspective of written sources, imperial governments presented a unified culture. Even local gazetteers, handbooks of local history and culture, were written by literati who identified with the central culture of whatever dynasty they lived in. While those literati listed the important history and culture of the county under consideration, they enumerated the aspects of a locality that were considered important by literati. Much like the evolution of theatre or drama, local culture was refashioned into elite culture by literati justifying local or popular culture in conventional Ru terms.

All of the three major strands of Chinese theatre developed in the second half of imperial history. Northern or Yuan songs (*zaju*) emerged in the thirteenth century and remained popular until the fifteenth. This partially overlapped with the southern tradition (*chuanqi*), which was popular from the fourteenth to the mid-seventeenth century. As this faded away, Peking opera (*jingju*) became important, though it didn't reach its mature form until around 1870. While northern songs and the southern tradition became literary arts, Peking opera was far less concerned with drama. All of which supports the idea of regional variation, but it is also true that local forms of theatre, north and south, interacted with and strongly influenced each other from as early as the eleventh century. Because the pool of subjects for theatre was firmly established by the Yuan, however, new works were well-known stories adapted into specific regional styles, rather than new tales. The audience knew these larger, familiar stories, making it unnecessary for performances to completely produce the entirety of a plot or provide background to particular scenes. Pieces, rather than complete works, of the resulting music dramas could be, and often were, performed by themselves along with pieces of other works. Some became perennial favorites and many of these have been reproduced in movies in the twenty-first century. Like Western operas, all dramas were musical, and all characters followed specific types, making the performance itself, and the adaptation of the story, the key determinant of quality for the audience.

Drama alternated between spoken and sung passages, with the rhymed singing, sometimes chanted and sometimes with music, broken up by acrobatic demonstrations. This development of prose and rhyme began with storytellers in the Tang dynasty. Storytelling followed a single narrative plot, introduced characters as they came up, and provided extensive descriptions of the background to the scenes. Drama continued these practices with respect to plot, and characters introduced themselves when they appeared on stage. The storytellers' use of description to establish setting translated into very minimal sets.

Growing wealth and urbanization in the Song dynasty laid the groundwork for the full development of drama. For the first time, permanent performance spaces were set up to provide regular entertainments. Theatres were established in the pleasure quarters, known as "tile markets." One explanation of the term "tile market" was that the performers and audience came together and dispersed as rapidly as roof tiles were assembled and disassembled, but it otherwise remains obscure. There were ten tile markets in the Northern Song capital, Kaifeng; twenty-three in the Southern Song capital, Hangzhou; and others in large cities and towns around China. The tile markets contained theatres, restaurants, shops, and brothels, and were in many cases open throughout the night. Many officials were troubled by these areas where elites and commoners rubbed shoulders, dissolute youth could get into trouble, and public entertainment was continuously on offer. Despite those concerns, and perhaps because the same officials were themselves visitors to those venues, the tile markets continued to flourish.

During the Song, individual tile markets had many theatres, offering many different kinds of performances. There were famous musicians, like Bai Mudan, poet comedians, like Zhang Shanren, and martial artists, like the wrestler Han Fu, among others. Martial arts practices became more flowery in the theatres of the tile markets, as they were performed alongside silhouette shows, and dramatizations of historical events. Like many modern audiences, people in the Song learned their history from popular renderings on stage. Performers were known for their specialized skill in one area, with Hou Sijiu noted for his stories of the Three Kingdoms period, and Yin Changmai for his tales of the Five Dynasties. Women and men performed, further troubling culturally conservative literati.

Performances were aimed at a broad audience so they were "vulgar" in comparison to the literary arts, which required a significant education to appreciate. Tile market entertainments were market driven, but much of the storytelling content, whether in prose or verse, was the same set of history, poetry, strange tales, and legends that had always been available to the educated. Performances in the tile market theatres transformed respectable high culture into popular low culture. Of course, the tile markets did not initiate this change, they just concentrated it in urban centers, where money and a larger population amplified and accelerated the rate of change. The innovation of the phenomena and its urban milieu also placed it directly under the scrutiny of literati, who then visited the tile markets and wrote about these new, popular entertainments.

Two particular forms, the All Keys and Modes, and Farce performances, contributed to later music drama. The All Keys and Modes form was a ballad carried by sets of songs and dialogue. Farce performances were either shorter sections within longer performances, or an entire comedic show in a four-act structure. The troupes that performed both forms maintained a set of standard roles that appeared in all of these shows. Both the four-act structure and the use of fixed roles by professional troupes carried over into the general development of theatre.

The northern or Yuan drama followed the four-act format, with a single narrative line, accompanying music, and only one singing character per play. With only four or five major scenes, it followed a regular pattern of introduction, elaboration, twist, and resolution. The second and third act were the major action and lyric parts, with the shorter fourth act confined to the restoration of order following the events of the play. Comedy, acrobatics, or martial-arts interludes broke up the dramatic exposition. Northern plays developed out of folk traditions of storytelling, resulting in a large range of topics. That said, the themes and topics of subsequent plays actually shrank over the later part of imperial history, as newer works focused exclusively on older, already-established topics.

Two factors led to the narrowing of dramatic topics. First, as literati became involved in writing plays, they imposed their own values on the form. As a group, the Ru values of the literati tended to frown upon salacious content, the endorsement of anti-social behavior, and anti-government sentiment. Vulgar folk entertainment of the commoners became literary art with political and cultural power. While some literati struggled to make drama a legitimate art, others, inside and outside government, now took it seriously enough to censor it. Second, in the Ming and Qing dynasties, the government restricted the content of plays, prohibiting lewd material, expressions of political opposition to the government, and stories about emperors.

Ming authors of northern plays were higher status literati than their Yuan predecessors. They also wrote southern plays, too, which were much freer musically, with multiple characters singing, the use of various musical styles (including northern music), and a different set of instruments. Like northern plays, the characters fit a standard set of types. Every southern play centers on a pair of lovers, regardless of the surrounding or underlying issues of the piece. Although southern plays had existed earlier, during the heyday of the northern play, they were the dominant dramatic form by the middle of the Ming dynasty. From the middle of the sixteenth century to the end of the seventeenth, literati playwrights composed a large corpus. This was not only part of the shift in interest among the literati, who now took drama seriously, but also of a shift to an alternative way of experiencing a play. Literati now began to read plays privately, rather than just see them performed. Booksellers, in response, began to produce illustrated scripts. Southern plays were still performed, by both professional troupes and amateurs, until well into the eighteenth century. Even in the nineteenth century, well after Peking Opera became dominant, they were still an important form. The Taiping Rebellion (see Chapter 6, p. 92), however, devastated the core area of southern drama, pushing it off the Chinese stage.

Peking Opera developed late in imperial history. All performances intersperse singing and acting, with "Civil" plays concentrating on emotions, often love, and interpersonal relationships, and "Martial" plays concentrating more on martial arts and acrobatic performances. The Qianlong emperor (r.1736–96) gets some credit for its development, because he watched various local operatic forms on his six southern tours taken between 1751 and 1784. On 25 September 1790, a local troupe from Anhui was invited to Beijing to perform for the eightieth birthday of the Qianlong emperor. The Four Great Anhui Troupes established themselves in the capital over the next few decades, and were then joined by troupes from Hubei, who spoke the same dialect of Chinese. In the middle of the nineteenth century, the combination of Anhui and Hubei forms produced Peking Opera (Beijing theatre style), which would go on to survive the fall of the Qing dynasty, still being seen as the classical form of Chinese opera today.

# Popular Literature

Most of the literature that is known in China was written down by literati, who rarely concerned themselves with the popular literature that was orally transmitted and performed. Many of these popular stories were adopted into the dramatic tradition, but storytelling existed long before plays, and continued through the end of imperial history. The great advantage of storytelling was that its performance required almost nothing, just a storyteller and an audience. From the historical perspective, however, the great disadvantage of those simple requirements was that they did not require a written record.

Storytelling may well have existed for centuries before the beginning of imperial China, but the first mentions of professional storytellers date from the Tang. Song sources provide considerably more detail, including the names of famous male and female storytellers, as well as their specialized repertoires. In the Southern Song capital of Hangzhou, stories purportedly fell into four groups, which one modern reconstruction describes as: historical stories of the rise and fall of dynasties, including the wars; the Song–Jin conflict (recent events); serious and comic religious stories; and general stories of romance, crime, corruption, and bandits. This tradition continued through the Yuan into the Ming and Qing dynasties.

Knowledge of the storytellers and their stories is confined to literati writings that happen to mention them, vastly underrepresenting storytelling, and its influence on drama and literature. Records state that performances fell into two categories, prose and chantefable (a story told in alternating prose and poetry). Prose stories were told by men, alone, without a written text. The content of these stories was broad, but usually quite familiar to the audience, and performed without anything apart from a fan and two pieces of wood to clap together. Chantefables often used written texts, male and female performers, and moved between prose and poetry, which was chanted or sung. Poetry could predominate in some stories, but it was not the sort of poetry that literati thought much of. The origins of the chantefable are unclear, with some claiming Indian origins—via Buddhist sutras, leading to Tang "parallel prose"—and others suggesting indigenous origins.

As a popular form, both prose and chantefable storytelling practice varied widely in different parts of China. The contents of the stories, however, were common across the empire, forming a reservoir of culture that literati drew upon when composing plays and literature. Certain cultural heroes like Guan Yu or Judge Bao (999–1062), who were righteous men, even in the face of corruption, featured regularly, a practice that continues into modern cinema. At the very end of the imperial era, some revolutionaries appropriated the pre-existing storytelling form to inject their new political and cultural ideas into the populace. Foreign literature introduced both new forms and new content to Chinese literature, along with Communism and a strong political awareness of the power of the existing social system to prevent change. In order to reach out to less-educated rural audiences, Communist performance troupes often staged revolutionary stories in traditional formats. This would reach its height during the Cultural Revolution (1966–76) when the only permitted theatre was the "Eight Model Operas."

# Conclusion

Considerable content flowed between the elite and popular arts in China. Local stories were recomposed by literati and written into dramas and literature. Aspects of culture as diverse as gardens and drama were accessible to commoners, even as they made claims to more exclusive elite tastes. Most storytelling was never regarded as "great" literature, and remained mere entertainment for the commoners, but educated men and women might still be in the audience to hear, and enjoy, those stories. The real break between the elite arts and common entertainment was literacy and education. It took a literate person to read a story or drama, and understand a piece of calligraphy. It took education to recognize allusions, hidden meanings, and the relationship of one piece of art to another. In all of this, the importance of the arts, both elite and popular, prompted the government and even private literati to insist upon censorship of the topics represented. The emperor could not be ridiculed or criticized, nor could art promote rebellion, "perverse" religious sects (those outside the established institutions), or the overturning of the social order. Those efforts, judging by the topics portrayed in all of these forms, were successful. Even works like *The Outlaws of the Marsh* (*Shuihu Zhuan*), which focused on a group of bandits, eventually bring the righteous men back under government control. Whatever their creativity, the arts were never fully free in imperial China.

# 10

# Constructing China Through History

History and historians played a central role in the intellectual and political life of imperial China. History permeated political and philosophical thinking long before the Qin unification, when the state of Qin conquered all of its rivals and established the first "emperor" in 221 BCE, and was fundamental to the construction of imperial dynastic legitimacy. Apart from serving as a record of the past, it had three main political functions. First, writing history established the legitimacy of the previous dynasty, a prerequisite to legitimizing the new ruler. Second, the historical reputation of a ruler was one of the only ways of limiting an emperor's absolute power. Court officials often suggested to a ruler that truly heinous acts would be remembered for centuries, marking him as a bad person. And, third, past events were genuinely useful in deciding on government policies. Since Chinese officials and rulers were usually well educated, they shared a similar knowledge and approach to history. In this sense, history also functioned as a medium of political communication that elevated the significance of history, historians, and the use of the past. Essentially, history was too important to be left only to private historians.

Although rulers and their dynasties claimed Heaven's Mandate as a divine sanction for their authority, a dynasty's political reality was established in military conquest and historical narration.

Heaven's Mandate moved from dynasty to dynasty, and a legitimate dynasty succeeded a previously legitimate dynasty by the transfer of the Mandate. By writing the history of the previous dynasty, one could establish its legitimacy and set foundations of one's own. This practice established the "Orthodox Succession (*zhengtong*)" of dynasties, a sequence that was often a subject of debate.

As had been the case before imperial times, the threat of history was one of the only restraints on a ruler's power. Chinese rulers were theoretically absolute, subject only to Heaven's displeasure, and the possibility of a tarnished legacy. While officials might use a natural disaster or unexpected heavenly event to argue that Heaven disapproved of a particular policy or behavior, for the most part invoking Heaven was a limited tool. A more promising method was to inculcate future emperors with the possibility that their deeds and misdeeds would be recounted long after they were gone. By constantly discussing the good and bad emperors of the past, the future emperor's teachers tried to encourage rulers to see themselves as part of a historical succession. One could choose to be a good emperor by emulating the positive models of the past, or neglect those models and be forever held up as a bad emperor.

History also provided a catalogue of policies and their outcomes for government statesmen and emperors to draw upon. Though of course, highly educated officials were capable of offering differing interpretations of the outcomes of policies, or providing counter examples. The larger challenge for policymakers was that a new policy that had no precedents was inherently problematic. This problem had its roots in Qin ideology, the Legalist underpinnings of which explicitly prohibited the use of the past to criticize the present. This was necessary because much of the Qin approach to governing was presented as a break with past practice. Had Qin rule and its Legalist ideology not become the paradigmatic example of how imperial government should not function, perhaps explicit innovation in government policy might have been acceptable in later dynasties. As it was, any new policies had to be described as consistent with the past, rather than as a break from it.

History writing was not just a political act, however, even if most historians wrote for the government or served as officials. Chinese historians were deeply concerned with the problems of their sources, and methods of analysis, in an effort to get to the truth. When it came to the role of the historian, the perennial question was whether they should comment on the events and people under discussion. A historian could bias history, or fulfill his didactic role by selecting events and never explicitly commenting on them. Master Meng claimed that Master Kong had edited *The Spring and Autumn Annals* (the earliest extant annalistic history, and one of the Ru classics), not only making one of the China's most important cultural figures a historian, but also allowing Master Kong subtly to "praise and blame" the actions of people in the past. By simply changing the verb used to describe a ruler's death, the historian could indicate whether he thought that ruler legitimate. A historian could also highlight negative or positive actions, juxtapose decisions and outcomes, and leave out events based on unreliable sources. These more and less subtle techniques had real-world consequences for the legitimacy of governments and political figures.

Once the Qin conquest created a government that unified temporal as well as spiritual control over the Chinese world, it became possible to write a coherent political history of that world from the perspective of a single government. The pre-imperial histories were either state histories, or idealized moral histories of the "spiritually," or perhaps more accurately, "heavenly" unified rules of the Zhou dynasty and, perhaps, the Shang and Xia dynasties. The latter was the approach of people like Master Kong and some other Warring States period thinkers, and it carried over into the imperial period. Imperial history was the history of a moral and temporal authority. From the imperial government's perspective, the writing of history that did not assert the primacy of the unified government of the Chinese world was an attack on dynastic legitimacy.

Because imperial governments supported history writing in some form, with later dynasties being more systematic than earlier ones, imperial history was maintained uninterrupted from the Qin to the fall of the Qing. That history was court-centered, seeing the world with imperial eyes, politically, militarily, and culturally. Thousands of official historians worked to produce those histories, most of whom are unknown to us. A very small number made major contributions, in particular the two great Han dynasty historians Sima Qian and Ban Gu, and the great Song historian Sima Guang. Later historians emulated their work, as they had emulated the work of their predecessors.

## Sima Qian (c.145/135–c.86 BCE) and *The Records of the Grand Historian*

Sima Qian was the most important historian of imperial China. Not only is his great work, *The Records of the Grand Historian* (*Shiji*), the main source for much of pre-imperial history, its form, consisting of Annals, Biographies, and Treatises, established the basic structure of all subsequent official dynastic histories. Sima's biographical categories were themselves highly influential, grouping individuals together as exemplars of their categories, and even creating the named category itself. His writing style also became a model for future generations. His personal story, and the extreme sacrifices he made to complete his history, further enhanced his reputation, as did the veiled criticisms he made of current events and people through the careful selection of the past events he included.

Sima Qian's father, Sima Tan, had been appointed as *Taishi*, an official position responsible for formulating the annual calendar and recording important events. The *Taishi* was thus both an astrologer/astronomer, since the calendrical calculations included determining auspicious and inauspicious days, and a historian or scribe. In English, *Taishi* has been variously translated as "Grand Historian," "Grand Scribe," or "Grand Astrologer," making the work Sima Qian would become famous for, the *Shiji*, into the [Grand] Historian/Scribe/Astrologer's Records. The term "*shi*" came to refer only to history and historians early in imperial history. Even so, astrology or astronomy were important parts of dynastic histories, and astral events were duly recorded. Some historians stressed the close connection between astral and political events, though others were skeptical.

Sima Tan began work on what would become *The Records of the Grand Historian*, but had not completed it when he fell ill and died. His model was *The Spring and Autumn Annals*, the history of the state of Lu (in modern Shandong) from 722–481 BCE that Master Meng claimed had been edited by Master Kong. This was the oldest Chinese history in annals form. It was also one of the Five Classics, a group of ancient texts that were considered canonical by the Ru before imperial times. In Sima Tan's dying charge to his son, he attributed *The Spring and Autumn Annals* to Master Kong, reiterating Master Meng's claim and further amplifying the significance of history writing. Sima Qian resolved to complete his father's work. After the standard three-year mourning period following the death of one's father, he succeeded to the position of *Taishi*.

This filial project was almost derailed by an unfortunate turn of events at court. The emperor's brother-in-law, Li Guangli, a mostly unsuccessful general, and Li Ling, a general whom Sima Qian respected, led an expedition against the Xiongnu, a nomadic steppe confederation in 99 BCE. The Xiongnu had been, and would continue to be, at odds with the Han dynasty as the two powers butted up against each other. In this case, the two Han generals were defeated and captured. Although the emperor and most of the court blamed the defeat on Li Ling, Sima defended him. Emperor Wu saw the defense of Li Ling as an accusation against Li Guangli, and ordered Sima castrated. While it was possible to buy off this punishment, Sima lacked the means to do so. Under such circumstances, it was expected that he would follow the honorable course of committing suicide, rather than suffer the disgrace of castration. But Sima could not leave *The Records of the Grand Historian* unfinished.

Not only did Sima survive castration and complete his work, but he also wrote a famous letter to his friend Ren An explaining why he followed that course of action. His father, he admitted, had not accomplished any great deeds, and his work on astronomy and the calendar was not very different from interacting with spirits or doing divination. The emperor had not taken him seriously, treating him like an entertainer in the court's retinue. Sima realized that if he killed himself, no one would care and he would be forgotten. Moreover, as far as anyone else was concerned, if he died it would have been because he was guilty of the crime. Dying was easier than living, but he would not be able to do something memorable and become known to future generations. Thus, following in the tradition of great men like Master Kong, Qu Yuan, and Master Sun, among others who, despite (or perhaps because of) desperate circumstances, produced texts to pass down, Sima chose to try to be known rather than forgotten. As it happened, he was so successful that he subsequently ranked with the great men he looked to for inspiration.

Before *The Records of the Grand Historian*, there were histories of individual states or their ruling houses. This regional and local diversity was characteristic of the flourishing culture of the Zhou dynasty, but also a cause of its military and political conflicts. As part of its conquest, the Qin actively destroyed the histories of those states, narrowing the knowledge of history to that of the

Qin, and the perspective of the Qin state. Despite these efforts, enough historical materials remained for Sima Tan and then Sima Qian to write a history that was not exclusively from the Qin perspective.

Nevertheless, the sources limited what Sima could write about, and most of his sources were lost soon after he used them. *The Records of the Grand Historian* was, and continues to be, the only account of much of what is known for parts of pre-imperial history. Until the twentieth century, when modern archaeology brought to light new materials, much of what Sima had to say about pre-imperial history could not be corroborated. Even what he had to say about the Qin was strongly biased by Han dynasty perspectives. Sima's personal biases also played a role, for example, as he projected his criticisms of Han emperor Wu onto the first Qin ruler. This was partly "using the past to criticize the present," the usual accusation against historians, and partly simply replacing the reality of a past figure with the characteristics of a living one.

Although *The Records of the Grand Historian* was widely read, and understood as fundamentally important, many critical historians argued over the real value of its contents; historians in imperial China were just as skeptical as modern historians. But Sima Qian's five sections—Basic Annals, Chronological Tables, Treatises, Hereditary Houses, and Biographies—undeniably established the building blocks for future history writing. All imperial histories would also include Basic Annals, Treatises, and Biographies. The Basic Annals was a chronological account of events from the perspective of the ruler (before the Qin), and the emperor afterward. This account assumed the centrality of the emperor and his court—with information and memorials being offered to him, officials debating policies before him, edicts being issued—but also featured the emperor at leisure, or performing the rites of government. The Treatises offered detailed discussions of more technical and institutional subjects like the rites, music, the calendar, astronomy, and the economy, among other issues. The Biographies section was broken up into categories like the Ru, Legalists, militarists, and so on. Each category then presented the biographies of men who were exemplars of that classification.

Subsequent dynastic histories dispensed with the Chronological Tables and Hereditary Houses, and provided different mixes of topics in the treatises and biographical categories, but the overall form was that initiated by Sima Qian. More powerfully, some of the categories in *The Records of the Grand Historian* grouped thinkers together, or separated them from one another, in ways that persist until the present. Sima's six "schools" of thought represented a particular Han-dynasty view of Warring States period thinkers rather than a contemporaneous view of them. Even there, he was simplifying the "Hundred Schools of Thought" of the Warring States period into a more understandable framework that highlighted what he perceived to be the major themes in that disparate proliferation of thinkers. Sima had similar effects on other aspects of later thinking about intellectual traditions, as with his highlighting of military thinkers, for example, which made Master Sun and Wu Qi into the pre-eminent military thinkers for the rest of Chinese history.

## Ban Gu (32–92) and *The History of the Han (Hanshu)*

The second great history written during the Han dynasty was Ban Gu's *The History of the Han*, which covered the period 206 BCE to 23 CE and would later be referred to as *The History of the Former Han* to distinguish it from Fan Ye's *History of the Later Han*, which covered the period from 6–189 CE. Like Sima Qian, Ban Gu took up his history writing from his father, Ban Biao (3–54 CE), whose unfinished history had been intended as a continuation of Sima Qian's work, which had only chronicled some of the early Han dynasty. Also, like Sima Qian, Ban Gu fell prey to political problems, demonstrating how important and dangerous history writing could be in imperial China. Ban Gu died or was killed before completing it, leaving it to his sister, Ban Zhao (48?–116? CE), to do so. The Ban family had a long tradition of scholarship and education that extended to the women in the family as well. Whether the Ban family's tradition of education was a rarity is unknown.

While Ban Gu was working on his father's incomplete history, someone informed the emperor what he was doing. Ban was arrested for compiling a dynastic history, put in prison in the capital, and had his books confiscated. Fortunately, his younger brother was able to get an audience with the emperor and explain what Ban Gu had been doing. This explanation, coupled with an examination of what had already been written, not only got him out of jail, it also got him appointed as an official historian with orders to complete his work. In all, it would take Ban twenty years to write *The History of the Han*, and its framework followed that set by Sima Qian. The form of the dynastic history was now functionally set.

## History Writing in the Tang Dynasty

The first mention of an official History Office was during the Northern Qi (550–77), and by the seventh century, during the Tang dynasty, a History Office in some form became a regular part of the imperial bureaucracy. Official Tang historians had compiled five dynastic histories by 636, historiographically legitimizing the dynasty and making sense of the chaotic period from the fall of the Han to the Tang founding. In 646, the second Tang emperor, Tang Taizong, ordered the compilation of *The History of the Jin*, and even personally wrote a few passages in it. Taizong was writing against the team of scholars employed to compose the history, who were using the opportunity to reiterate their advice to the emperor that he not place imperial clansmen in government positions or indulge in military adventures. Where the scholars saw an opportunity for remonstrance, the emperor sought an example of imperial unity, since the Jin had unified China during its rule from 265 to 420.

While *The History of the Jin* was an example of conflict between an emperor and his officials, it was more likely that political factions within the government would fight over the historical record. Once it became standard practice to compile documents and outline histories of current events, the various living political actors all had a stake in how they were portrayed. When a different faction got into power, they revised the historical record to vilify their enemies and valorize themselves. After a few changes of power, it was almost impossible to salvage a less politicized record of events. Even without political interference, the process of editing all of the administrative activity around an emperor into a manageable form required extensive cutting and compression. By the twelfth century, officials recorded full accounts of the emperor's daily activities and discussions. These records were stored, edited, and then summarized. Officials themselves kept court diaries in addition to copies of their memorials that were also supposed to be provided to the History Office. On top of that, separate compilations of critical documents (*huiyao*) were compiled to inform the later writing of Treatises on institutions.

Institutional history writing as a separate subject took a huge step forward during the Tang. Although Sima Qian and Ban Gu had established the form of the dynastic history, there was still room for new forms of history outside that paradigm. As previously discussed, dynastic histories came to affirm the legitimacy of the dynasty the historians lived in by legitimizing its predecessor. By choosing a set of exemplary generals and statesmen, or paradigmatic rebels, a historian asserted mainstream values and established who was right or wrong in the historical ledger. The basic annals presented the emperor at the center of the world. The Treatises section was somewhat underdeveloped in political terms; as a mostly technical discussion of important institutional functions, it seemed merely explanatory.

The use of institutional history for political purposes changed with Du You's compilation of *The Comprehensive Institutions* (*Tongdian*) in 801. Du had served as prime minister on three occasions, and so was intimately familiar with the functioning of imperial government. His innovation in writing a purely institutional history was twofold: First, it argued for dynastic legitimacy based upon effective government function and, second, that a dynasty was its institutions. The first argument justified the second, and the second was necessary because Du You sought to bolster the legitimacy of the Tang dynasty after the An Lushan Rebellion had almost ended it. Of course, Du had to assert in the introduction that effective institutions accorded with the Ru ideology of caring for the people, rather than the Legalist vision of building up the government to empower the ruler.

## History Writing in the Song Dynasty

History writing in the Song dynasty continued many Tang developments even as it added new and more sophisticated practices of its own. After producing imperially sponsored histories of the Tang and Five Dynasties periods in the late tenth century, the early emperors also ordered the creation of encyclopedias (*leishu*) that ranged broadly over literature, the classics, and history. The inclusion of both Daoist and Buddhist perspectives, in addition to Ruist perspectives, showed the openness of Song emperors in the early Song dynasty. Although those non-Ru perspectives continued, the main currents of history writing were firmly Ruist, however, and often actively hostile to Daoist and Buddhist influence.

The first important historical shift in the eleventh century was the privately produced *Historical Records of the Five Dynasties* (*Wudai Shiji*) by Ouyang Xiu (1007–72). Ouyang was a prominent official and intellectual, who sought to rewrite the pre-existing *History of the Five Dynasties* (*Wudai Shi*) into a more didactic, less documentary history. Instead of including extensive quotes of memorials, Ouyang wrote an elegant, condensed rendering of events that made their moral lessons clear. To further insure the transmission of his message, he inserted more extensive comments explaining his position than any previous historian had done. The *Historical Records of the Five Dynasties* was not widely circulated, for unknown reasons, and was only made generally available after Ouyang's death at the court's insistence.

Ouyang Xiu was also involved in the rewriting of the *History of the Tang* (*Tangshu*), which was finished in 1060. Both of these new histories were revised to stress praise and blame, and to make them more relevant to current eleventh-century events. Ouyang's explicitly moral position coupled with a superb writing style made his histories clearer and more compelling. They were also considerably shorter than the prior histories of those periods. The lessons of history, or at least Ouyang Xiu's lessons, could be quickly absorbed by any educated person studying for a civil service exam. Ultimately, these two works would displace the older ones, with Ouyang's histories being referred to as *The New History of the Five Dynasties* (*Xin Wudai Shi*) and *The New Tang History* (*Xin Tangshu*).

However, the greatest monument to history writing in the Song was written by Sima Guang (1019–86), supported by a team of historians and scribes. Sima's *The Comprehensive Mirror for Aid in Governing* was also a partly private work, covering the period from 403 BCE to 959 CE, in 294 chapters. Although the emperor sponsored Sima's project, providing funding, personnel, and supplies, it was compiled and written outside the official government History Office. Sima Guang, it should be noted, was a senior government official, and would become chief councilor shortly before his death, so he was by no means an unorthodox or anti-institutional figure. Indeed, he was an arch conservative in both his historical and political practices. Where Ouyang Xiu emphasized clear moral messages, Sima emphasized chronology and sources. One of his great innovations was to include the "considerations of differences" (*Kaoyi*), where conflicting sources were presented in the text along with the chosen account of what happened. Like Ouyang Xiu, Sima included non-official sources for history (private histories, literary collections, etc., all produced outside government bureaus), but he applied a critical apparatus to those, as he did for all sources, not only making transparent his history-writing process, but also helping the reader to develop their historical judgement.

Although Sima's work was created for the use of the emperor, it enjoyed much wider circulation. The main difficulty with *The Comprehensive Mirror*, as Sima himself acknowledged in a preface, was that it was so long and complicated that everyone fell asleep after a few pages. With events described in chronological order, something that turned out to be significant much later might be long forgotten by the time a reader reached the climax of a particular story. This was realistic, in the sense that what was and was not significant in the past might not be known until much later, but it required the reader to keep seemingly minor events in their head in case they mattered later on. Where Ouyang was an idealistic classicist who believed that history proved the morals expressed in the classics correct, Sima was a generally realistic historian who believed that there were authentic lessons to be learned from the past. Just as good people could improve the world by working to do good, so too could bad people harm the world by doing bad things. The lesson for the emperor was that he had to choose good policies and good ministers or the empire would fall into ruin and the dynasty collapse.

*The Comprehensive Mirror* was admired by many literati, but it was simply too cumbersome and its message too morally subtle

government officials who at least began their work outside the official, government writing of history. Sima Guang's intended audience was clear, first Emperor Yingzong and then, after his death, Emperor Shenzong. Sima Qian sought to be remembered and hopefully appreciated by an "extraordinary" person in the future. Ban Gu was continuing his father's work, but probably only expected to be read by men like himself. All three had profound impacts on the educated class, and through them the culture as a whole. These exemplars, along with the thousands of other historians working in imperial China, not only produced (as officials) primary source documents, they also compiled and digested those documents into formal histories. Those accounts were not primary sources for the past, they were the products of history writing and professional historians. Most of what comes down to us as the annals of imperial China is the product of imperial Chinese officials, and therefore our understanding of the imperial past is based on the values and goals of those historians as reflected in their histories.

# 11

# The End of Imperial China?

A Chinese aphorism holds that "History and Literature cannot be separated." This is not an argument for writing history stylishly, but rather an acknowledgement that the construction of a historical narrative involves determining what to include, who is important, and what the point of that history is. A history of imperial China begins from a set of political choices that sets a clear time frame, and then moves to a set of cultural choices that, in fact, blur the territorial unit of analysis. Even the understanding of what Chinese culture is or was, or who counted as Chinese, is unclear. In the past and in the present, historians made their determinations based upon the requirements of government or their own historiographical needs, but in making those determinations none can stand outside the broad platform of Chinese history itself. The writing of Chinese history, even in English, is also part of Chinese history.

While there was continual change in imperial Chinese history, in cultural and historiographical terms the two most important periods were the first century of the Han dynasty, in roughly the second century BCE, and the high point of the Northern Song dynasty, during the eleventh century. Scholars and statesmen in the early Han dynasty laid the foundation for many aspects of what became imperial ideology and culture by consolidating an understanding of pre-imperial times, and codifying the changes brought about by the Qin and Han dynasties in breaking with that earlier period. A similar process took place in the eleventh century when the strands of thought and culture from the preceding millennium were gathered together, digested (particularly in the literary sense of producing anthologies), and reimagined into a new foundation for the succeeding millennium of imperial Chinese history. The temptation in looking back at this history from the twenty-first-century perspective is to ignore both the continual changes and the critical periods of change in favor of a uniform, static narrative that reflected or established what China was, and what it meant and what it means to be Chinese in the present.

# Imperial History

History has always been a tool of Chinese governments, something that was explicitly understood even before the imperial period. That use was occasionally contested by individual historians in the imperial period, but the ideological and philosophical orientation of classical Chinese education strongly biased the majority in favor of the conventional and orthodox, leaving only a few true rebels against the established order. Of that small group of rebels, most still sought to displace orthodoxy with their own ideas and become, in turn, orthodox and legitimized by the government. This was particularly true of scholars who were more interested in using history to support their larger intellectual programs than in actually undermining conventional historiography itself. The category of imperial history is now an active area of struggle for all of the traditional reasons that Chinese governments have always sought to control the narrative: The past justifies the present.

The same can be said of the physical domain of empire, with earlier imagined boundaries legitimizing later claims. But the territory of imperial China was never stable, whether across dynasties, or even within a single dynasty. Imperial dynasties were not modern nation states with clearly delineated borders, or clearly established populations. Dynasties usually claimed direct authority over more territory than they actually controlled, and asserted that Heaven's Son was sovereign of All-Under-Heaven. This expansive claim to power over virtually everything was tempered by an acknowledged limit over concentric zones of diminishing real control. Still, rulers from those outer zones beyond the direct control of the emperor were supposed to acknowledge the ritual sovereignty of Heaven's Son over their territory. In reality, of course, distant rulers were out of the emperor's reach and their only concessions to imperial ideology would have been in diplomatic protocol.

Another difficulty in defining imperial Chinese history is the question of who counted as Chinese or, at least, as part of imperial society. Many groups of people living near or within the borders of a given dynasty identified themselves as something other than Chinese. Some of those groups, like the Mongols and Manchus, actually created imperial governments that controlled most or all of the Chinese ecumene. Neither the Mongols nor the Manchus defined themselves as Chinese, and they treated the Chinese population as one of a number of distinct groups under their control. Most of the history of these non-Chinese participants in Chinese history was, however, written in Literary Chinese within the conventions of Chinese historiography. While there were rulers who had their own written languages, most of the bureaucratic documents of imperial governments were written in Literary Chinese. This was true even when the emperor did not speak, read, or write any form of the language.

## Diversity

The dominance of Literary Chinese throughout the historical record should not obfuscate the diversity of lands, cultures, and peoples that have been forced into the category of imperial China. Yet that category remains as much a political as a historiographical question. Dynasties claimed lands and peoples, and asserted control over orthodox culture, by establishing or re-establishing imperial power. Those institutions were created and maintained by force, and then written into the historical record. Even diversity had a place in that record, if only as a negative example of uncivilized, anti-social, or heterodox behavior. Truly disruptive concepts were suppressed by not being recorded at all. This was easy enough to accomplish given how little was ever really recorded, even in the historically rich culture of imperial China.

Diversity was more obvious at the local and regional levels. Imperial governments were defined by the centralization of authority, in theory if not always in practice. The Qin unification of China was based upon Heaven's Son holding both ritual and temporal power over the empire. Regional diversity in coinage, writing, and other aspects of culture (including the histories of individual states) was stamped out, and a single coinage, writing style, and culture imposed everywhere. Diversity of any kind was antithetical to the power of the imperial center, even if local culture, from religion to food, always flourished. Imperial orthodoxy was not a complete fiction, every dynasty imposed real values on the people and territory under its authority, but it was never the totality that imperial officials maintained.

The growing pool of educated men and greater wealth led to local, and therefore diverse, centers of culture and power. Writing local histories became increasingly popular over the last millennium of imperial history, as did activism in local society. Wealthy families understood that their long-term goals were best served by creating and maintaining a strong, local base of power. The ideal of serving in government remained, but the likelihood of passing the civil service exams, the main route to becoming a civil official, was so low as to seem a question of luck or divine intervention rather than education. Thus, educated men found alternative arenas in which to use their training, and those pursuits weakened their commitment to the imperial government, which struggled to maintain the loyalty of a gentry class that could not realistically expect much from the central authorities.

## Unity

While there were always centrifugal forces tearing at the cohesion of imperial governments, there were also centripetal forces maintaining peace and order. The unifying grasp of the written Chinese language (keeping in mind the wide diversity of different local spoken languages) cannot be easily ignored. Elites shared a powerful, deep literate culture that tied them not only to one another, but also to an imagined shared history. That literate past was often directly connected to geography, emphasizing the Yellow River and the Central Plains as the core of Chinese culture. More than anything else, the flexibility of the connections between culture, politics, and geography allowed for a sense of continuity across time and place. China was not an exact thing that could be grasped or lost, it was a firm core that retained its existence even in the face of political turmoil, geographic division, or cultural schisms.

Imperial rites and ceremonies tried to impose, or assert, unity, but common religious beliefs were just as important in supporting the idea of a unified culture across the empire. Some local spirits or cults might have been confined to small areas, but the Three Teachings spanned the empire and persisted during periods of disunion. Even the divisions implicit in speaking of the Three Teachings obfuscates the larger ideal of a shared truth that all three traditions sought. Buddhism, Daoism, and Ruism agreed to disagree with one another, while utterly rejecting heterodox cults that undermined the moral order. All three remained in competition with one another throughout imperial history, though all had also become entirely mainstream. More generally, a shared set of festivals and holidays spoke to a unified culture and people who celebrated a series of annual activities regardless of place or politics.

Of course, those who did not share those annual holidays, or maintained a different set, could clearly be identified as not Chinese.

On a more prosaic level, the imperial economy became increasingly integrated through trade over two millennia. Economic progress was not linear, or even consistently positive, but the transport networks and fiscal systems of the empire did connect provinces and disparate, distant places. China was too large, of course, to describe the resulting economy as a single market. Economic development was very different in different parts of the empire, with some areas achieving levels comparable to those of the most advanced parts of Europe in the early modern period, and others remaining backward. Indeed, despite extended periods of political unity, and the formation of a homogeneous political ideology that spanned the empire, economic development and culture was highly variable. That variability frequently introduced tension into the imperial system.

The power of imperial Chinese culture created ritual, political, and economic ties across a vast, diverse landscape. That unifying culture subsumed the variations caused by environment and lifestyle, allowing people to identify as Chinese while also retaining their individual ties to their community. Since the majority of imperial subjects within the ecumene saw themselves as Chinese, the presence of non-Chinese minorities did not disrupt imperial culture. Indeed, there were non-Chinese imperial subjects who saw themselves as part of a dynasty without becoming Chinese. There were also non-Chinese rulers, and non-Chinese ruling classes, like the Mongols or Manchus, who could rule China without becoming Chinese. Imperial culture, rather than Chinese culture, unified the empire, even if the imperial culture was Chinese.

# Conclusion

China has never been a single place, or a single culture, but from 221 BCE to 1912 CE the idea prevailed that an individual,

Heaven's Son, should rule over All-Under-Heaven because he—and with one notable exception it was always a he—held Heaven's Mandate. The authority of Heaven's Son was theoretically absolute, but he ruled only so long as his benevolent policies benefited the people as a whole. Misrule would lead to the loss of Heaven's Mandate and, consequently, to the loss of one's throne. Imperial rule was fundamentally based upon the idea that a government was supposed to take care of the people under its authority. And Heaven's Son held authority over everyone.

This theory of government made no mention of nationality, ethnicity, or race since, of course, no such thing existed at that time in the modern sense of those terms. Barbarians were people who had different customs, beliefs, and languages from the Chinese, but they were welcome to take up Chinese practice, and those who did so were the most praiseworthy. China's culture was open in the sense that it was a set of practices and beliefs anyone could adopt. It was also open insofar as it absorbed the influences of many other cultures, from the Buddhist religion to New World chili peppers. Today it is hard to imagine Chinese food without New World crops, but those arrived fairly late in imperial history. Similarly, many attitudes and perspectives from the last dynasties, some coming from Western views of China, were not present for most of the imperial period. Outside ideas and commodities became a part of Chinese culture, which was always in a state of flux.

The moments of greatest change were, of course, imperial China's beginning and its end. In 221 BCE China was unified under the rule of a single emperor and on February 12, 1912, the last Qing emperor abdicated. The latter was a political shift that dramatically changed many aspects of Chinese culture and initiated decades of war. The values of imperial China, based firmly in a textual tradition written in Classical and Literary Chinese, were set aside, but that they did not entirely die out, and may even revive in some form, is a testament to their enduring value.

# Acknowledgments

I would like to thank Tony Stewart and Robert Campany, two of my colleagues at Vanderbilt University, for their kind advice on the topic of religion, and my editor, Jonathan Bentley-Smith, who has spoiled me for any future projects. I would also like to thank my wife, Tracy, and my daughters, Aileen and Lindsay, for being who they are and bringing joy into my life.

# Further Reading

Brook, Timothy, *The Troubled Empire: China in the Yuan and Ming Dynasties*, Cambridge, MA: Belknap Press, 2013.

Cai, Zong-qi (ed.), *How to Read Chinese Poetry in Context: Poetic Culture from Antiquity Through the Tang*, New York: Columbia University Press, 2018.

Clunas, Craig, *Art in China*, 2nd edn, Oxford: Oxford University Press, 2009.

Denecke, Wiebke, Li, Wai-Yee, and Tian, Xiaofei, *The Oxford Handbook of Classical Chinese Literature (1000 BCE–900 CE)*, Oxford: Oxford University Press, 2017.

Ebrey, Patricia Buckley, *The Inner Quarters: Marriage and the Lives of Chinese Women in the Sung Period*, Berkeley and Los Angeles: University of California Press, 1993.

Ebrey, Patricia Buckley, *Women and the Family in Chinese History*, London and New York: Routledge, 2002.

Graff, David A., *Medieval Chinese Warfare, 300–900*, London and New York: Routledge, 2002.

Hsiao, Kung-chuan, *A History of Chinese Political Thought, Volume One: From the Beginnings to the Sixth Century A.D.*, trans. F. W. Mote, Princeton: Princeton University Press, 1979.

Kim, Youngmin, *A History of Chinese Political Thought*, Cambridge, UK, and Medford, MA: Polity Press, 2017.

Knight, Sabina, *Chinese Literature: A Very Short Introduction*, Oxford: Oxford University Press, 2012.

Kuhn, Dieter, *The Age of Confucian Rule: The Song Transformation of China*, Cambridge, MA: Belknap Press, 2011.

Lewis, Mark Edward, *The Early Chinese Empires: Qin and Han*, Cambridge, MA: Belknap Press, 2010.

Lewis, Mark Edward, *China Between Empires: The Northern and Southern Dynasties*, Cambridge, MA: Belknap Press, 2011.

Lewis, Mark Edward, *China's Cosmopolitan Empire: The Tang Dynasty*, Cambridge, MA: Belknap Press, 2012.

Lorge, Peter, *Chinese Martial Arts: From Antiquity to the Twenty-First Century*, Cambridge, UK: Cambridge University Press, 2012.

Mair, Victor H. (ed.), *The Shorter Columbia Anthology of Traditional Chinese Literature*, New York: Columbia University Press, 2000.

Rawson, Jessica (ed.), *The British Museum Book of Chinese Art*, London: British Museum Press, 1992.

Rowe, William T., *China's Last Empire: The Great Qing*, Cambridge, MA: Belknap Press, 2012.

Steinhardt, Nancy, *Chinese Architecture: A History*, Princeton: Princeton University Press, 2019.

# Index